Manufacturing technology volume 1

KT-385-438

G. Kenlay
Senior lecturer in production engineering,
Nene College, Northampton

and

K. W. Harris
Lecturer in mechanical engineering,
Northampton College of Further Education

Edward Arnold

© G. Kenlay, K. W. Harris 1979

First published 1979
by Edward Arnold (Publishers) Ltd
41 Bedford Square, London WC1B 3DQ

All Rights Reserved. No part of this publication may be reproduced, stored
in a retrieval system, or transmitted in any form or by any means, electronic,
mechanical, photocopying, recording or otherwise, without the prior
permission of Edward Arnold (Publishers) Ltd.

British Library Cataloguing in Publication Data

Kenlay, G.
  Manufacturing technology.
  Vol. 1
  1. Production engineering
  I. Title   II. Harris, K. W.
  621.7   TS176

  ISBN 0–7131–3401–1

Text set in 10/11 pt IBM Press Roman, printed and bound
in Great Britain at The Pitman Press, Bath

Contents

Preface   iv

1 **Primary forming processes**   1
Introduction. The production of iron and steel. Sand casting.
Rolling. Drawing. Extrusion, Forging. Relative merits and
limitations of the primary forming processes

2 **Simple presswork**   18
Costs of production. Presswork and its uses. The fly press. The
power press. Press tools. Press-tool operations. Presswork planning.
Safety

3 **Metal-cutting and cutting fluids**   33
Metal-cutting tools. Analysis of metal-cutting. Positive- and
negative-rake cutting. Cutting forces. Factors affecting the
cutting process. Cutting-tool materials. Functions of a cutting
fluid. Types of cutting fluid. Application of cutting fluids

4 **Machine tools**   54
Elements of a machine tool. Degrees of freedom. Generating,
forming, and copying. Capstan and turret lathes. The drilling
machine. The shaping machine. The grinding machine. The
milling machine

5 **Measurement**   74
The purpose of measurement. Measurement of length. Flatness.
Straightness. Squareness. Roundness. Accuracy of measurement.
Comparative measurement. Measurement of angles. Measuring
'theoretical dimensions'

6 **Heat treatment**   108
Introduction. Constituents and composition of carbon steels.
Annealing. Normalising. Hardening. Tempering. Case-hardening,
Materials testing. Plain-carbon steels. Selection of heat-treatment
processes. Furnaces. Choice of heating medium. Measurement of
furnace temperature. Quenching. Safety

7 **Plastics**   132
The uses and nature of plastics. Types of plastics. Plastics-forming
processes. Advantages and limitations of moulding processes. The
use of inserts. Choice of material and process. Safety

Index   153

Preface

This book has been specifically written to cover the requirements of the Technician Education Council standard unit Manufacturing technology II (U76/056) for technician students of mechanical and production engineering.

Our aim has been to explain, in simple concise terms, the general and specific objectives listed under each of the topic areas in the standard unit, using many large diagrams to keep the descriptive material, and the text itself, to a reasonable length. The appropriate British Standards have been referred to throughout the book, and the student is encouraged to refer to them for further information if required.

We are indebted to many of our colleagues for help and encouragement while writing the book, but our special thanks must go to Mr R. C. Smith of Nene College for the many hours spent proof-reading and to Bob Davenport of Edward Arnold (Publishers) Ltd for his help and patience throughout the long time we have spent preparing the manuscript.

Finally, extracts from British Standards are reproduced by kind permission of the British Standards Institution, 2 Park Street, London W1A 2BS, from whom copies of the complete standards may be obtained; figs 1.1, 1.2, and 1.3 are based on material in the British Steel Corporation booklet. *Making steel*, by kind permission; fig. 5.2 is reproduced by courtesy of TI Coventry Gauge Ltd; and fig. 5.37 (crown copyright) is reproduced by permission of the Controller of Her Majesty's Stationery Office.

G. Kenlay
K. W. Harris

1   Primary forming processes

1.1 Introduction

Primary forming processes are processes which produce either materials suitable for subsequent processing or components requiring machining before they are ready for use.

There is an ever increasing range of engineering materials in many different forms of supply now available to the engineer, and the task of selecting a particular form of material supply to fulfil a specific engineering function, consistent with economic considerations, is becoming increasingly difficult. In addition, there are also a growing number of processes used to convert materials from their primary form (i.e. the form in which they are supplied ready for making things) to finished manufactured articles.

In this chapter we shall outline the ways in which materials are obtained in their primary forms and describe some of the more basic processes used in industry to produce engineering components.

The forming of metals is one of the oldest 'production' processes known to man, and for many years metals such as gold, bronze, and iron have been formed into useful objects by the techniques of casting and forging. The technology of these processes has obviously developed over the years, but, nevertheless, their fundamentals have remained unchanged. Mineral ores are mined from the earth's crust and are heated to convert the ore into a raw material. The raw material may then be cast either directly to the shape required or as a lump of metal that may be worked into shape at a later stage.

In this book we shall deal only with the metals iron and steel, as these are by far the most widely used metals in any industrial society, but in chapter 7 we shall also discuss the forming of plastics, which are becoming increasingly important.

1.2 The production of iron and steel

Iron and steel are obtained from iron ore, which is found in many parts of the world. It is an earthy rock, consisting of up to 70 % iron. Unfortunately, the iron is in the form of an iron oxide from which the iron has to be extracted. This is done by a process known as *smelting*, in which heat is applied to the ore in a blast furnace, to bring about a chemical reaction.

The blast furnace

The blast furnace is used for the first stage of the conversion of iron ore into the many forms of ferrous (iron-containing) materials available to the engineer. It produces a material called *pig iron.*

The production of pig iron in the blast furnace, fig. 1.1, is a continuous process. The 'charge', consisting of iron ore, coke, and limestone, together with 'sinter' (fine particles of iron ore, coke, and limestone roasted in a sinter

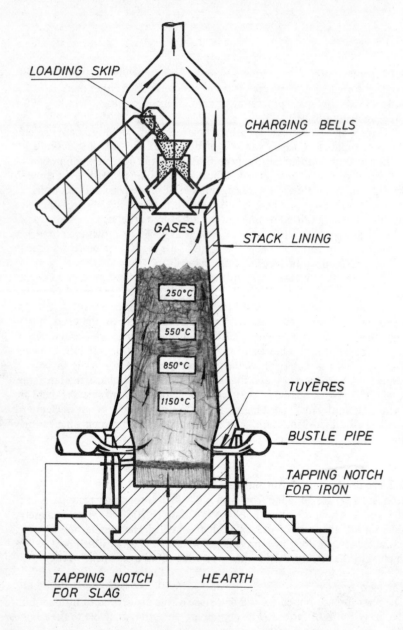

Fig. 1.1 The blast furnace

plant to form a clinker rich in iron), is conveyed to the top of the furnace and is fed in through a system of hoppers called *charging bells*.

The heat and the carbon required to enable the smelting process to take place inside the furnace are produced by the coke and a blast of hot air which is blown into the furnace via the bustle pipe and a series of nozzles, called *tuyères*, placed around the bottom of the furnace. The oxygen in the air causes the coke to burn fiercely and provide the heat and the reducing gases that extract the iron from its earthy impurities. The molten iron produced gradually collects in the *hearth* at the bottom of the furnace.

The limestone in the charge helps to separate the earthy impurities from the ore and forms a molten slag that floats on top of the molten iron. When sufficient iron has collected in the hearth, the level of the slag rises to the *slag notch*, which is normally sealed with fire-clay. The fire-clay is then removed and the molten slag is tapped off into ladles. The molten iron is removed through a tap hole at the bottom of the furnace. In the past the iron was run into open sand moulds known as pig beds, so called because they resembled piglets feeding from their mothers — hence the name 'pig iron'.

Pig iron

Pig iron from the modern blast furnace is used in two ways. If the blast furnace is part of a steelworks then the pig iron is run into a large ladle called a *receiver* and is transferred in the liquid state to steel-making furnaces. The liquid pig iron, known as 'hot metal', is then refined without it being necessary to reheat it, which means a great saving in cost.

There are several different types of steel-making furnaces in use in the steel-making industry, producing a wide variety of steels for specific purposes. We shall not deal with the different types of steel-making furnace in this book — it is sufficient to say that the steel produced is cast either into large masses known as *ingots*, weighing up to several tonnes, or into *slabs* and *billets* by a fairly new process called *continuous casting*. Slab and billet casters produce a long continuous length of the appropriate section by allowing the liquid steel to solidify in a water-cooled mould. As the steel becomes solid it is drawn from the bottom of the mould, allowing more metal to solidify on top of it; hence a continuous ribbon of white-hot steel is obtained, fig. 1.2. This 'ribbon' is cropped into convenient lengths which are transported to a rolling mill for further processing.

Where the pig iron is not to be converted into steel then invariably it will be cast into solid 'pigs' in a pig-casting machine. This is simply an endless chain carrying metal moulds which are filled with molten pig iron and cooled with a water spray. The machine is usually arranged so that the solid pigs are automatically dumped into waiting railway waggons. The pig iron is then transported to iron foundries in various industrial areas to be converted into cast irons.

Pig iron will vary in composition depending upon the type of iron ore used. The carbon content is generally between 3 % and 4 %, with varying amounts of silicon, sulphur, phosphorus, and manganese.

3

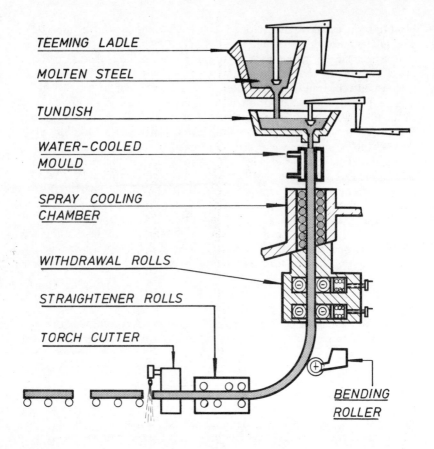

TEEMING LADLE

MOLTEN STEEL

TUNDISH

WATER-COOLED MOULD

SPRAY COOLING CHAMBER

WITHDRAWAL ROLLS

STRAIGHTENER ROLLS

TORCH CUTTER

BENDING ROLLER

Fig. 1.2 Continuous casting

Basic raw-material forms such as ingots and solid pig iron have no direct uses as such — further processes must be carried out on these materials to convert them into forms more familiar and useful to the engineer. Processes such as *casting, rolling, drawing, extrusion*, and *forging* are typical, and these are known as primary forming processes.

1.3 Sand casting

Cast iron
Most sand castings are produced from cast iron, which is a ferrous material produced in a furnace called a *cupola*, fig. 1.3. Cast iron has a relatively low melting temperature, is free-flowing in the molten state, basically brittle and weak in tension, but strong in compression. Cast irons generally contain

4

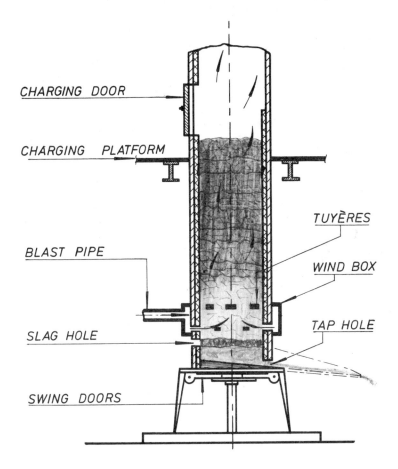

Fig. 1.3 The cupola

between 2 % and 4 % carbon, 1 % to 3 % silicon, and smaller amounts of sulphur and phosphorus.

To produce cast iron, the cupola is loaded with pig iron, coke, limestone, and steel scrap. The charge is then subjected to a reduction process similar to that of the blast furnace, although the production of cast iron is not a continuous one like the blast furnace. When the cast iron is molten, it is tapped off at the bottom of the furnace into ladles. The ladles are then used to pour the cast iron into previously prepared moulds. When solid, the cast iron thus takes up the shape of the cavity inside the mould.

Basic steps of sand casting
The first step in the process of sand casting is the production of a *pattern* to produce the mould cavity. The pattern is made by a skilled craftsman, usually

from wood, although it may be made from metal (for durability) or plastics (for cheapness). Patterns are made slightly larger than the required casting, to allow for the contraction of the molten metal that takes place on cooling. Patterns are also invariably made in more than one piece, to facilitate the removal of the pattern from the mould and help minimise any damage to the mould that may be caused during this skilled operation.

The simple casting shown in fig. 1.4 requires a two-piece split pattern and a two-piece *moulding box*, sometimes called a *moulding flask*. To prepare the mould required to produce such a casting, half the pattern is placed on a *turnover board* and the bottom half of the moulding box, called the *drag*, is positioned around it. The half pattern is then covered with a *facing sand* – a special mixture of sand and clay that will withstand the high temperatures involved without breaking down and giving a poor surface to the finished casting. The drag is then filled with *green sand* (a sand containing moisture and clay), which is rammed up and levelled off. The ramming allows the drag to be turned over without the sand falling out. When this has been done, the top half of the moulding box, called the *cope*, is placed on the drag and is aligned by means of locating pins. The other half of the pattern is then located by its dowels, and a dry layer of *parting sand* is sprinkled over the green-sand surface to enable the two halves of the moulding box to be easily separated. A similar procedure with facing sand and green sand is then carried out.

If a casting is required to have a hole or some other form of internal cavity (as in the case shown), then provision for a *core* must be made. The core will require its own moulding box and is usually made from a sand containing a binding agent, so that the core has sufficient strength when finished to be able to support itself in the mould. It is located in what are known as *core prints* – impressions left in the sand by additional protuberances on the pattern.

Provision has to be made for filling the cavity with molten metal and allowing the air and gases generated during the pouring process to escape. For this, tapered wooden *runner* and *riser* plugs are appropriately positioned during the ramming of the cope. The riser plug (or plugs, depending on the size and shape of the casting) are placed at the highest part of the mould. The runner plug is usually placed to one side of the cavity – this is so that an *in-gate* (or in-gates) can be cut when the cope and drag have been separated. This technique avoids hot metal dropping directly into the cavity, causing damage and/or displacement of any cores.

When the two halves of the mould have been separated, the pattern is removed by screwing a steel spike into it and tapping gently in all directions – this process, called *rapping*, assists in the removal of the pattern from the mould and helps minimise any damage that may be caused. The finishing touches can now be completed, such as the cutting of the in-gate in the drag and the forming of a pouring cup in the cope.

The pouring cup is a depression into which the molten metal is poured, so that it overflows and runs down the runner into the mould cavity rather than dropping in from a height. The mould also has to be 'vented' by poking a wire into the cavity to produce tiny outlets for the gases generated when

6

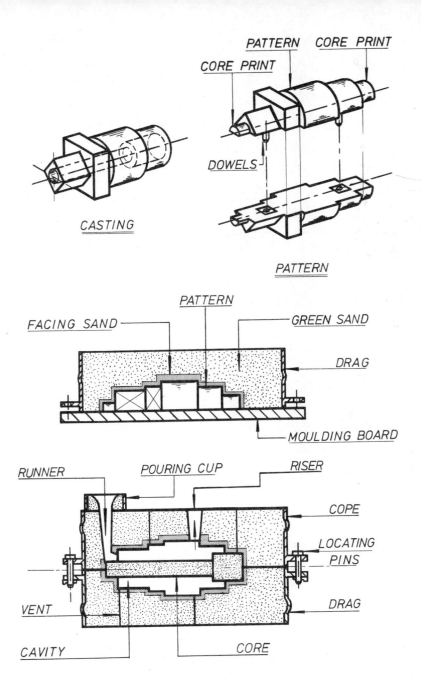

CASTING

CORE PRINT

PATTERN CORE PRINT

CORE PRINT

DOWELS

PATTERN

FACING SAND

PATTERN

GREEN SAND

DRAG

MOULDING BOARD

RUNNER

POURING CUP

RISER

COPE

LOCATING
PINS

VENT

DRAG

CAVITY

CORE

Fig. 1.4 Basic steps of casting

the hot metal comes into contact with the damp sand. Any loose sand is blown out of the cavities, which are then coated with blacklead, which helps give the casting a good surface finish. If the mould requires a core, this is carefully located in the core prints and the two halves of the moulding box are then assembled and relocated in exact alignment by the pins.

The molten cast iron is then carefully poured until the mould is filled up to the top of the runner and riser openings. Although the metal in the runner and riser has to be removed when the casting has solidified, it serves a useful purpose in that it acts as a reservoir to feed metal back to the casting as it contracts on cooling. When the casting has solidified, the mould is broken away and the casting is vibrated to remove any loose sand. The runners and risers are broken off with a hammer, and any rough surfaces are *fettled*, i.e. rough ground to remove the sharp edges.

The foundry thus supplies the machine shop with sound clean castings of a desired shape, ready for marking out (if required) and machining.

1.4 Rolling

The majority of steel products start their life by being converted from the ingot or continuous-cast form by a process of rolling, although some large forgings are produced directly from cast steel ingots. (The process of forging will be described later.)

If the steel has not already been cast in the slab or billet form by the continuous-casting process but is supplied as ingots, then its preliminary treatment will consist of reducing its section by rolling in what is known as a *two-high reversing mill*, fig. 1.5. As its name suggests, this mill rolls the ingot one way, reverses its two main rollers, and passes it back in the other direction. These mills are of two basic types – *cogging* mills and *slabbing* mills. Cogging mills produce *blooms* which are about 150 mm square in section, or larger, and several metres in length. Slabbing mills produce

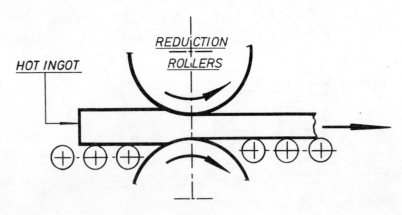

Fig. 1.5 A two-high reversing mill

8

rectangular slabs of between 75 mm and 250 mm thick and 300 mm to 2000 mm wide.

Various types of rolling mill are used to process slabs and blooms still further into a range of products designed to suit industrial needs, fig. 1.6. Slabs form the basis for the production of plate, sheet, and coiled wide strip.

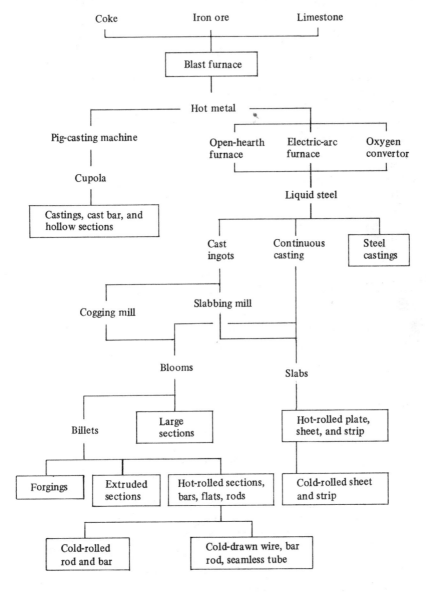

Fig. 1.6 Flow chart showing iron and steel products

Blooms are used for a range of intermediate products such as billets, large rounds, small slabs, and heavy sections (girders and structural sections) from which a great variety of smaller sections are then rolled, such as bars, flats, rounds, sheet, and light sections (angle-iron channel, RSJ's, etc.).

These products are worked hot, as less energy is needed to deform metal at high temperature. This is because steel when heated gains in the property of *malleability* (the ability to be squeezed or hammered by compression without fracture) at the expense of *ductility* (the ability to be drawn out in tension). Unfortunately, the hot working causes the surface of the steel to oxidise and scale, giving it a blue-black appearance, and pickling in a weak acid solution has to be carried out to remove the scale and produce a surface that is acceptable. Also, the accuracy of size of a hot-worked material is more difficult to control than that of one cold worked.

There are methods of improving the surface finish and appearance of hot-rolled materials by working them in the cold state. By so doing, the accuracy of the product is greatly increased and, in addition, the *work-hardening* that takes place during the cold working improves the mechanical properties. Cold rolling is a typical example and is widely used to finish hot-rolled sheet, strip, and bars to produce bright-finished products of accurate size.

1.5 Drawing

Drawing is a cold-working process and consists simply of pulling the material to be drawn through a hole in a hard steel block, called a *die*, in order to reduce its diameter. To achieve this, the end of the bar is reduced to enable it to be passed through the hole in the die. It is then attached to a powered draw-bench which pulls it through the die, reducing its diameter and increasing its length, fig. 1.7.

The material being drawn down needs to be ductile, in other words it must be able to be stretched without fracture — lead, for instance, would not be suitable for drawing as it would break under the force required to reduce its

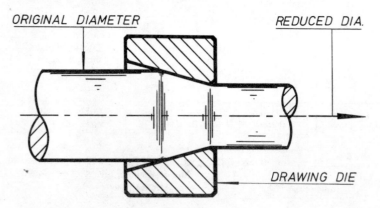

ORIGINAL DIAMETER   REDUCED DIA.

DRAWING DIE

Fig. 1.7 Fundamentals of cold drawing

diameter. Due to the mechanical deformation that takes place when drawing, even ductile materials work-harden. If large reductions in diameter are required, involving passes through several dies, then an inter-stage annealing operation (see section 6.3) would be required to restore the resulting loss of ductility due to work-hardening.

The die obviously needs to be extremely hard, to be able to withstand the abrasion that takes place between the material being drawn and the die. To reduce friction, the bar being drawn down is first passed through a box containing soap flakes or powder, which acts as a lubricant. The dies are made from such materials as tungsten carbide, hardened alloy steels, and diamond, depending upon individual requirements.

Drawing is used to produce bright-drawn bars and solid-drawn tubes, and is the main process of manufacture for the production of wire.

1.6 Extrusion

As a primary forming process, extrusion is a technique whereby a billet of *hot* metal is placed in the container of an extrusion press and is subjected to high pressure which causes the metal to flow plastically through a die placed in the end of the container, fig. 1.8. (Plasticity is the opposite of elasticity, i.e. after plastic flow the metal does not revert to its original shape.) It is a process that can be likened to squeezing toothpaste from its tube.

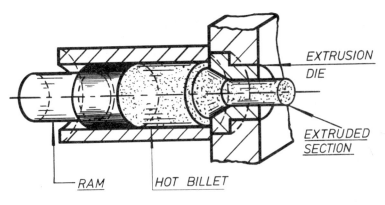

Fig. 1.8 Fundamentals of extrusion

The shape of the extruded product will depend upon the shape of the die, and many shapes of both simple and complex section, fig. 1.9, are now produced in long lengths (10–12 metres) in a single operation. It is also possible to extrude tubes and hollow sections.

Many metals can be successfully extruded, including copper, aluminium, magnesium, lead, zinc, steel, and their alloys. The temperature at which extrusion takes place varies: for steels it is between 1100 °C and 1250 °C, for brasses between 700 °C and 800 °C, and for aluminium alloys between 350 °C and 500 °C.

11

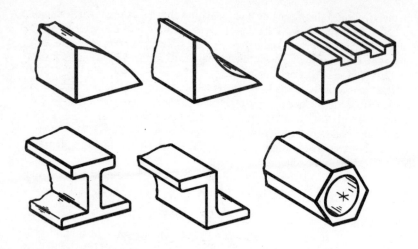

Fig. 1.9 Typical extruded sections

There are also cold-working extrusion processes, but these usually relate to the manufacture of *finished* articles such as small containers used for film, tablets, etc. and collapsible tubes used for such things as toothpaste, and so they are outside the scope of this book.

1.7 Forging

In the context of primary forming, forging is the term used for a wide range of similar processes which involve the deforming of a piece of hot metal to a desired shape by the action of a compressive force. The force may be of an impact type, like the blow from a hammer, or a squeeze, as with a hydraulic press. The force may be applied by hand or machine; to a single article, using simple hand tools, or to many thousands of components, such as the connecting rods used in car engines, which are produced with the aid of expensive dies and machines applying great force.

The principle behind all forging operations is the same, however — a bar of metal is heated to the correct forging temperature and a compressive force is applied to alter its shape. The advantages of this process are that, although the bar has to be heated and shaped, less metal will be used than if the shape were machined from a solid block of metal. Also, the grain flow of a forged component is usually considered to be an advantage as far as the strength and toughness of a component are concerned, fig. 1.10.

We will not deal with all the different forging processes, but will outline in sketch form the basic operations of hand forging, fig. 1.11. With all of these, the first stage, of course, is to heat the material to be forged to the correct temperature, which for plain-carbon steel will depend upon the carbon content (see fig. 6.1).

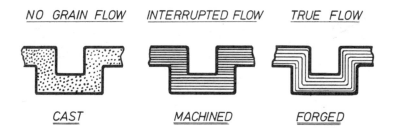

NO GRAIN FLOW   INTERRUPTED FLOW   TRUE FLOW

CAST   MACHINED   FORGED

Fig. 1.10 Grain flow

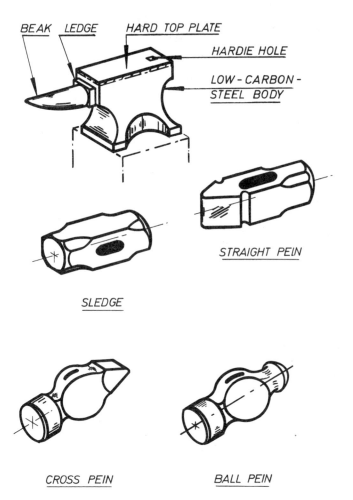

BEAK LEDGE   HARD TOP PLATE

HARDIE HOLE

LOW-CARBON-STEEL BODY

STRAIGHT PEIN

SLEDGE

CROSS PEIN   BALL PEIN

Fig. 1.11 Hand-forging tools and operations (*cont'd on pages 14 and 15*)

13

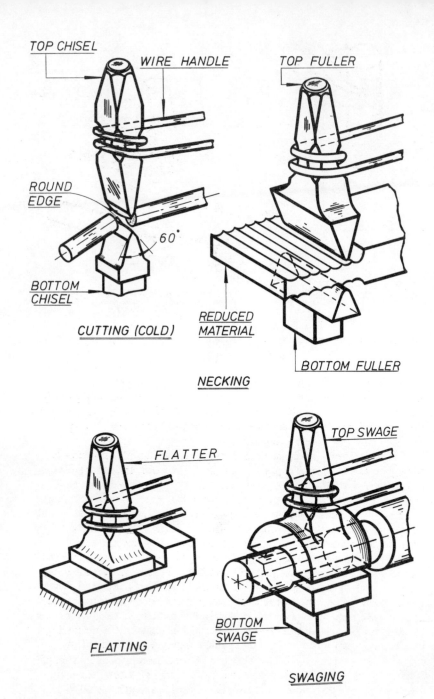

Fig. 1.11 Hand-forging tools and operations (*cont'd*)

14

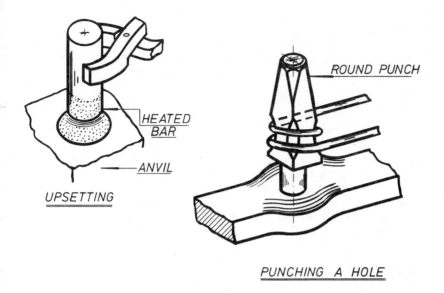

HEATED BAR

ANVIL

UPSETTING

ROUND PUNCH

PUNCHING A HOLE

Fig. 1.11 Hand-forging tools and operations (*cont'd*)

1.8 Relative merits and limitations of the primary forming processes

Although the processes described in this chapter have a common beginning — that of iron ore and the blast furnace — as a result of the different treatments the pig iron from the blast furnace receives, the end products of these processes will vary widely in character.

Cast irons, for instance, are fluid in the molten state and it is relatively straightforward to obtain complicated shapes by casting, though the cost of producing the pattern and the mould should be borne in mind. Cast iron, however, tends to be a brittle material — weak in tension but strong in compression — thus components produced from cast iron tend to need to be bulky and are used where they will not be subjected to high stresses. Typical examples are machine bases, cylinder blocks, gear-box housings, etc.

Steel, on the other hand, tends to be strong both in tension and in compression. It is a very versatile material, and, for the production of small parts, machining direct from regular-section material is usually economical. Where the shape and size of the part required would mean removing a large amount of material by machining (wasted material), it is sometimes more economical to heat a billet of metal and forge it to shape, thus reducing the machining time and effecting a cost saving on the amount of material required. As mentioned in section 1.7, there is also an additional benefit of extra strength due to the improved grain flow obtained by the forging process. The main disadvantage with forging is, of course, the cost of heating the metal to be forged and the need for forging equipment and a suitable size of furnace to achieve the forging temperature.

Obviously, the above factors have to be carefully considered when deciding whether a part should be machined from the solid bar or billet of metal, or whether it would be more economical to cast or forge it to shape. An internal-combustion-engine connection rod is a very good example of where a forging will save machining time and material, whereas the cylinder block of the same engine would probably be very expensive if produced by any process other than casting.

The processes of drawing and extrusion tend to be limited to the production of long continuous lengths of a particular cross-section which would be difficult to achieve by any other method. Drawing is usually carried out on metal in the cold state; hence this process is usually confined to ductile materials. Extrusion, on the other hand, is usually carried out hot, as the pressure required to force the metal through the die would be too great if the metal were not heated to its plastic state. Extruded aluminium bars are cut and mitred to form the frames of double-glazed windows, whereas the copper tube used for plumbing purposes is cold drawn.

1.9 Hazards

Extra safety precautions must be observed when in hazardous areas such as foundries and forges. Obviously routine regulations must be rigidly observed, but it is perhaps relevant to highlight here the dangers of hot metal. There are no second chances with molten metal, and carelessness causes many accidents each year.

Any form of moisture is extremely dangerous where molten metal is concerned. In the case of casting, for instance, where the moulding sand is moist, unless the steam generated is allowed to escape, violent explosions will occur and shower molten metal over a wide area.

Metal that has solidified and cooled to below red heat can still be as hot as $400\,^\circ C$ — four times hotter than boiling water — and will result in severe burns if it is picked up inadvertantly. At these high temperatures, scale forms on the surface of the steel and this, too, can be dangerously hot when removed by chipping or if it is cooled incorrectly. Thus care has to be taken in any area where there is hot metal, and in such environments it is essential that goggles, leather aprons, gloves, and steel-capped boots are worn.

Exercises on chapter 1

1 Name the materials used for the production of pig iron in the blast furnace. State the purpose of each of the materials used in the charge.
2 Describe the procedure for removing pig iron from the blast furnace.
3 For what purpose is pig iron used (a) in the solid form and (b) in the liquid form?
4 Molten steel from the steel-making furnace is cast into three different forms. What are they?
5 Describe the process of producing cast iron.
6 Describe with suitable illustrations the basic steps of producing a sand casting.
7 Explain how and why a sand core is used in the casting process.

8   Steel products such as angle iron, girders, bars, tubes, etc. delivered to engineering factories and building sites etc. have undergone many changes of shape before reaching their final form, and most of these changes have been carried out while the steel was hot. State why this is so and name the disadvantages of working metal hot.

9   List five types of tool used for hand-forging operations and explain the purpose of each tool.

10   State the advantages and disadvantages of the forging process.

11   Make a neat sketch to illustrate the principle of extrusion. Name six types of material that may be successfully extruded.

12   Describe the metal-forming operation known as drawing. For what purpose is this primarily used?

13   List the dangers that are predominant in areas working with hot metal or molten metal.

14   State what is meant by the term 'primary forming process'.

15   State the main advantages of the continuous-casting process.

16   Explain the meaning of the following: (a) cope, (b) drag, (c) green sand, (d) parting sand, (e) core prints, (f) runners and risers, (g) rapping.

17   For what reasons are slabs and blooms produced in a steel-making plant?

18   What are the main mechanical properties of a material that is suitable for drawing operations? Explain why these properties are important to the success of the drawing operation.

19   What is the main property of a material that is required to be extruded and why is this property important?

20   State the primary processes used in the production of (a) a bench-vice body, (b) an open-ended spanner, (c) a centre-lathe leadscrew, (d) a three-jaw self-centring lathe-chuck body, (e) a lathe bed, (f) a vee block.

2　Simple presswork

2.1 Costs of production

There are many expenses involved in the production of any component, and these include the cost of the materials used, the cost of the tools used, and the wages paid to the workers producing the component, as well as such things as the costs of the building in which the work goes on, power, administration, and so on.

The costs incurred in the production of a single example of a particular component are known as *unit costs* for that component. For example, *unit material cost* is the cost of the material (including wastage) required to make a single component; *unit labour cost* is the cost of the wages paid for making a single component (this may be a piecework rate for the component or the hourly rate paid to workers producing this component divided by the number of components they make per hour). Also, a particular tool may be able to produce only a certain number of components before it is worn out or until demand for that component ceases, so it is possible to calculate a *unit tooling cost:*

$$\text{unit tooling cost} = \frac{\text{cost of tool}}{\text{number of components to be made by tool}}$$

Costs such as these have to be taken into account when considering the production of any component.

2.2 Presswork and its uses

Presswork is a method of production involving the cold working of metals (and many non-metals), usually in the form of thin strip or sheet. By the very nature of the process, it is economical only for the production of large quantities of identical components. This is because each component produced has to have its own specially designed and manufactured press tool, and the cost of a particular press tool has to be recovered by spreading it over the number of components that the tool will eventually produce. For example, a press tool costing £3000 may be required to produce 100 000 components; the tooling cost per component thus works out at 3p, i.e. £3000 ÷ 100 000.

As presses can produce components at fairly fast rates (typical example, 60 small components per minute), the unit cost of labour for operating the press is very low. This makes presswork an ideal process for the low-cost production of many parts required in large quantities by modern industry.

When production of a particular component is required, the press tool is set-up in the press by a skilled setter. There are many different designs of presses in industry, but basically the function of any press is to provide a force to the ram of the press. This force may be applied by manual, mechanical, pneumatic, or hydraulic means. Here we shall deal only with the manually operated fly press and the mechanical power press.

2.3 The fly press

The fly press is the commonest of the manually operated presses. It is a bench-mounted press and its general features are shown in fig. 2.1. In use, the operator swings the handle, turning the multi-start screw and thus providing the up-and-down movement to the ram. The amount of force available to perform useful operations depends upon the amount of energy supplied by the operator. The masses on the arm act like a flywheel, storing up this energy which is then dissipated as the tools do their work.

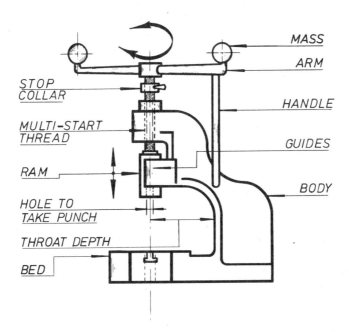

Fig. 2.1 General features of a fly press

Fly presses are used for the production of a wide range of small components requiring cutting operations (shearing of the metal) such as piercing and blanking or non-cutting operations such as bending. These operations will be dealt with in detail later, but many other operations, such as assembling parts together by riveting and crimping, can also be carried out.

The fly press is a relatively low-cost machine, generally using simple and inexpensive tools. No power supply is required, as the operator supplies the energy needed, but obviously the size of the component and the amount of force available to do useful work are limited. It is also relatively slow in operation, and operator fatigue has to be taken into account.

Many fly presses in industry are being converted to small power presses by removing the swinging arm and screw and replacing them with a pneumatic cylinder.

2.4 The power press

BS 4640: 1970, 'Classification of metalworking machine tools by type', classifies the power press as a *metal-forming* machine tool. With the mechanical power press, an electric motor supplies energy to a flywheel. When required, this energy is transferred via a clutch mechanism and a crankshaft to provide movement and force to the ram of the press, figs 2.2 and 2.3. Power presses are classified according to the amount of force that can be applied to the ram of the press and by the general features of their design, e.g. the type of frame — open-front C-type frame or double-sided.

The top half of the press tool is attached to the ram of the press and the bottom half is secured to the bedplate, fig. 2.3.

The variety of work that can be produced with power presses is enormous and, with good tool design, accurate production at low cost can be achieved. When one batch of components is finished, the change-over time, i.e. the time taken to remove the press tool and replace it with another, can often be less

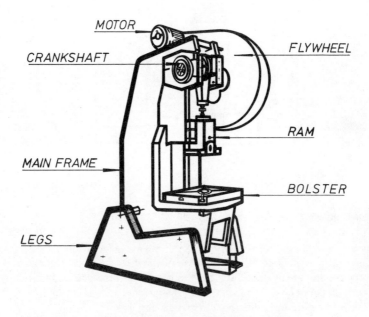

Fig. 2.2 A typical power press (guard not shown)

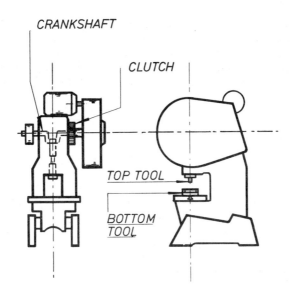

CRANKSHAFT

CLUTCH

TOP TOOL

BOTTOM TOOL

Fig. 2.3 Typical press tool positioned in a press (guard not shown)

than an hour — this ensures that the press itself, an expensive machine tool, is economically utilised, i.e. is not idle for too long. Although the press is a costly machine tool, its fast rate of operation enables the high tooling costs to be quickly recovered. Once set-up it can use unskilled labour for a production run, which again keeps the unit cost of a component low.

2.5 Press tools

Press tools may be broadly divided into cutting tools and non-cutting tools. As previously stated, each press tool has to be specially designed for a particular component; however, there are certain recognised elements of a press tool that can be itemised and illustrated.

A press tool may contain the elements shown in fig. 2.4. Not all press tools will have all these elements, and some tools will have additional features.

Punches and dies   These are the most important elements in any press tool, and their design has to be carefully considered. The size and shape of these parts control the shape and accuracy of the finished component.

Punches and dies have to be hard and strong, in order to produce many thousands of components; they are therefore made from a good-quality high-carbon, high-speed, or alloy steel, suitably hardened and tempered. When piercing holes of small diameter, the thickness of the work material must be taken into account, as the force required to produce these holes could possibly break the punch. A slender punch can be strengthened either by

21

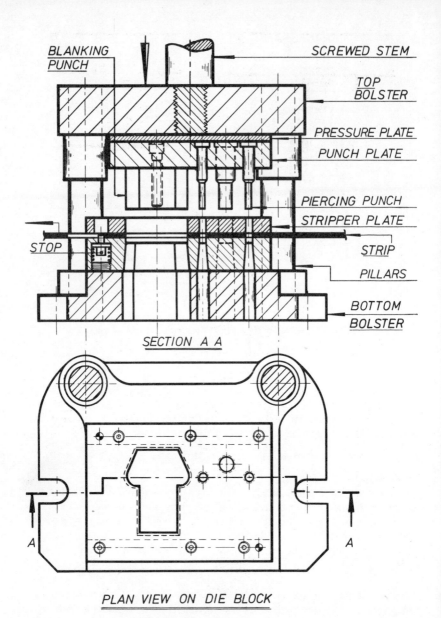

BLANKING PUNCH

SCREWED STEM

TOP BOLSTER

PRESSURE PLATE

PUNCH PLATE

PIERCING PUNCH

STRIPPER PLATE

STOP

STRIP

PILLARS

BOTTOM BOLSTER

SECTION A A

PLAN VIEW ON DIE BLOCK

Fig. 2.4 Typical press tool (follow-on type)

increasing the diameter in stages or by being supported with a 'quill'. Figure 2.5 shows typical punch-design features.

If several holes or blanks are to be produced at one stroke of the press, then the peak load acting on the punches can be reduced in two ways. The

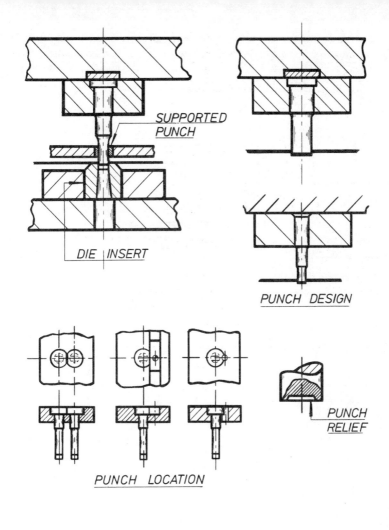

SUPPORTED PUNCH

DIE INSERT

PUNCH DESIGN

PUNCH LOCATION

PUNCH RELIEF

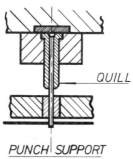

QUILL

PUNCH SUPPORT

Fig. 2.5 Design of punches

punches can be staggered, so that they do not all come into contact with the work at the same instant, or an angle known as 'shear' can be ground on the punch or the die, fig. 2.6. In order to keep the component flat, the shear is put on the punch when piercing and on the die when blanking.

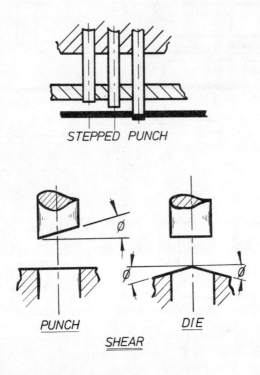

STEPPED PUNCH

PUNCH   DIE

SHEAR

Fig. 2.6 Staggering of punches and the use of 'shear'

Clearance   When used with reference to press tools, clearance is the space between the punch and the die, fig. 2.7. This distance is important in obtaining a good edge on the component, which obviously affects the accuracy of the part. The clearance is varied according to the thickness and hardness of the work material, and is usually expressed as a percentage of the material thickness, Table 2.1. Too large a clearance produces a component with ragged and burred edges, whereas too small a clearance gives rise to rapid punch and die wear and an increase in the force required to produce the part.

Top bolster (fig. 2.4)   This is a heavy slab of cast iron or low-carbon steel used for mounting the punches and the punch pressure plate. It also carries the screwed stem which enables the connection to the ram of the press to be made.

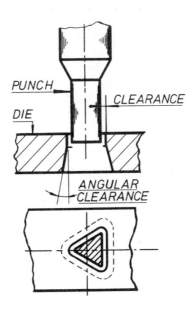

Fig. 2.7 Illustration of clearance

Material	% clearance
Low-carbon steel	5–10
Medium-carbon steel	6
High-carbon steel	10
Brass	6
Copper	6–8
Aluminium	7–10

Table 2.1 Press-tool clearances for different materials

Bottom bolster (fig. 2.4) This is similar to the top bolster but carries the die block and, where necessary, other features such as the stripper plate and stops. It enables the press tool to be securely fixed in position to the bedplate of the press, fig. 2.3.

Guide pillars (fig. 2.4) These keep the top and bottom halves of the tool accurately aligned, to enable tools used for cutting operations to maintain an accurate clearance between punch and die.

2.6 Press-tool operations

With presswork, the sequence of operations has to be decided upon early (there may only be one operation or there may be several). This enables the drawing office to design and the tool room to produce the press tools that will be required when production begins.

There are many recognised press operations, but, as this chapter is limited to simple presswork, only the operations of piercing, blanking, and bending will be considered.

Piercing This is a shearing operation used to produce a hole by means of a press tool. The piece removed is the scrap, and it is the hole size that is important. To produce an accurate hole, the punch is made to the size required and the clearance is added to the die size, fig. 2.8

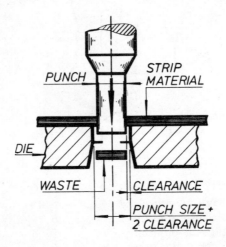

Fig. 2.8 Piercing operation

Blanking This is the operation of using a press tool to cut a shape from a strip or sheet, fig. 2.9. In order to obtain a blank size accurately, the die is made to size and the clearance is subtracted from the punch size.

Bending Bending is one of the non-cutting operations performed on a press. It uses a press tool to deform a component permanently to a required shape and is employed on many components such as trays, boxes, lugs, brackets and clips, etc. A typical punch and die used to bend a blanked component are shown in fig. 2.10.

When a metal is bent, the outer surface is stretched and the inner surface is compressed, fig. 2.11. The radius of a corner must therefore not be too small, fig. 2.12, otherwise too much stress will be set up and the material will crack on the outside.

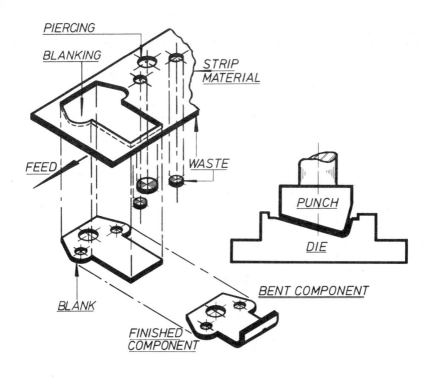

Fig. 2.9 Blanking operation   **Fig. 2.10** Bending operation

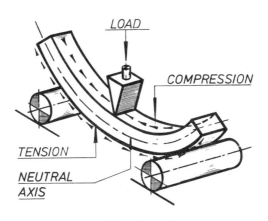

Fig. 2.11 Effect of bending metal

27

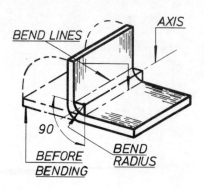

Fig. 2.12 Bend radius in relation to grain flow

The direction of the grain flow of the material in relation to the axis of the bend is also a factor to be considered. The most suitable condition is when the axis of the bend is perpendicular to the grain flow, fig. 2.13. As the angle α shown in fig. 2.13 decreases, so does the ability of a material to withstand severe bending.

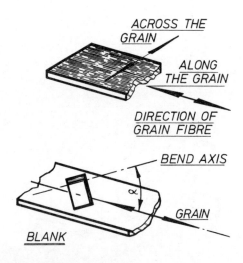

Fig. 2.13 Bend axis in relation to grain flow

When calculating the length of material required for a component that is to be bent, due consideration must be given to the stretching and compressing of the surfaces. For moderate bends it is usual to determine the length required by calculating the length of the neutral axis – the intermediate layer of material that is neither stretched nor compressed, fig. 2.11. Although in practice the position of the neutral axis varies for different materials, for our purpose it can be assumed to be midway, i.e. along the centre line of the component.

Example Calculate the length of material required to produce the pipe clip shown in fig. 2.14.

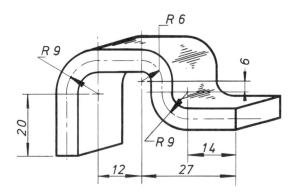

Fig. 2.14 Pipe-clip bracket

$$\text{Length required for straight parts} = 20 + 12 + 6 + 14$$

$$= 52 \text{ mm}$$

$$\text{Length required for curved parts} = \left(2 \times \frac{\pi 18}{4}\right) + \left(\frac{\pi 12}{4}\right)$$

$$= 9\pi + 3\pi$$

$$= 37.7 \text{ mm}$$

∴ Total length required = 52 + 37.7 = 89.7 mm, say 90 mm

2.7 Presswork planning

When planning presswork, it is necessary to determine a sequence of operations. Many factors have to be taken into account, the aim being to produce maximum quality of the product at minimum cost. There are no hard and fast rules; each component has to be studied individually to decide how it is to be produced.

The component shown in fig. 2.9 could have all its operations carried out singly using separate tools, e.g. blanking, piercing, and bending. The press tool shown in fig. 2.4. has been designed to pierce and blank at one stroke of the press, and is known as a *follow-on tool*. The component would then be completed by a second operation to bend it to the required shape.

The process of planning the most economical method of producing a part on a press is not easy, and experience is required to be able to do it successfully. The total number of parts required, the rate at which they are needed, and the type of presses that are available will all affect the type of press tool designed.

Blanking layouts
It has already been stated that the high production rates of presses using unskilled labour means that the unit labour cost of producing a component is low. It also means that the cost of the material itself is an important part of the total cost. As the cost of the material is directly related to the percentage of waste, the best layout of the component shape to give economy in material usage must be decided before considering the detail design of the press tool.

If the blanking layout in fig. 2.15 is examined, it can be seen that a single component is produced at each stroke of the press from strip material little wider than the component itself. If, however, it is arranged for the pressing to be produced every other space, then, by feeding the material through the press tool again in the turned-over position, further components are produced and the material waste is reduced by about 20%, fig. 2.16. Taking this a stage further, by using wider strip material and doubling the number of punches and die cavities, two components can be produced for each stroke of the press, with even better material utilisation, fig. 2.17. However, the extent to which it is possible to adopt these procedures depends upon the total quantity of parts required and the rates at which they are needed. Obviously the cost of the tool to produce two components per stroke of press would be greater, but the production time would be halved, and, as stated, there would be a small saving on material cost.

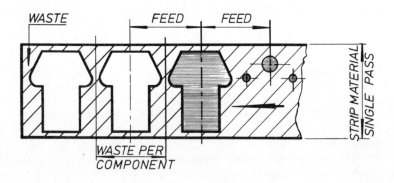

Fig. 2.15 Typical blanking layout

30

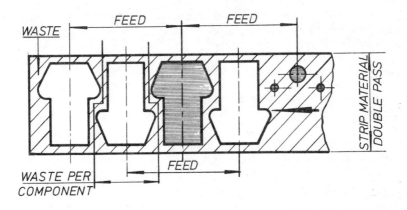

Fig. 2.16 Typical blanking layout

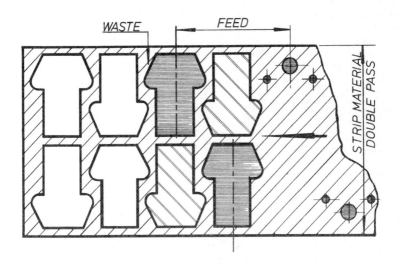

Fig. 2.17 Typical blanking layout

2.8 Safety

With all types of presswork, safety regulations are absolutely essential. The Power Press Regulations 1966 specify in great detail the guarding and operating conditions of all power presses. There must be no possibility of the operator's limbs being able to enter the tooling area, and a rigid system of inspection and authorisation that the safety appliances are in good working order is demanded, ensuring that a press is in safe working order at all times

Exercises on chapter 2

1 (a) Explain the features which make the fly press suitable for producing small components in small batch quantities.
b) Name six important parts of a fly press and describe their function.
2 (a) Describe, with the aid of suitable sketches, how power from an electric motor is transmitted to a press tool on a power press.
b) Sketch and describe the essential elements in a follow-on type of press tool.
c) Show three methods of retaining a punch in a press tool and explain why punches often vary in length when two or more are assembled in a press tool.
3 State three different metal-forming or metal-cutting operations that can be performed by the use of a power press.
4 When a component has to be bent to shape following a blanking operation, it is important that it is cut from the strip material in a given direction. Explain why this is so, and describe the likely effect on the component when the direction of cut is incorrect.
5 Make a neat sketch showing the relative positions of the punch, material, and die when piercing holes in strip material. Show the clearance and clearance angle and explain what the result would be (a) with insuffucient clearance and (b) with excessive clearance.
6 State two methods of ensuring that the top and the bottom halves are accurately aligned when setting up a press tool on a press.
7 When producing components from strip material, the width of the material is often governed by the shape of the component and the numbers required. Explain why this is so.
8 Make a sketch of a suitable tool to bend the component shown in fig. 2.18.
9 Calculate the length of material required to produce the component shown in fig. 2.18. [*Answer* 166 mm]
10 Explain why it is usually considered to be economical to produce components on a press only in fairly large quantities.

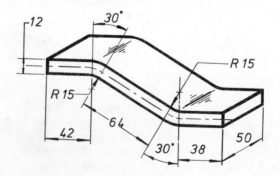

Fig. 2.18

32

3   Metal-cutting and cutting fluids

3.1 Metal-cutting tools

Metal-cutting is one of the most important processes which takes place in industry. The cutting action of all metal-removal tools is basically the same — be they chisels, hacksaws, or lathe or shaper tools — though some tools have many cutting edges and others have just one. It is important to consider what happens during the metal-cutting process, so that the most economical use can be made of available facilities.

Single-point tools

A single-point tool is classified according to BS 1296: Part 2: 1972 'Single-point cutting tools', as a tool terminating in a single cutting point. The cutting part is defined as the functional part of the tool, comprised of the chip-producing elements, fig. 3.1. The *major cutting edge* removes the bulk of the work material, and the *minor cutting edge* controls the surface finish of the workpiece.

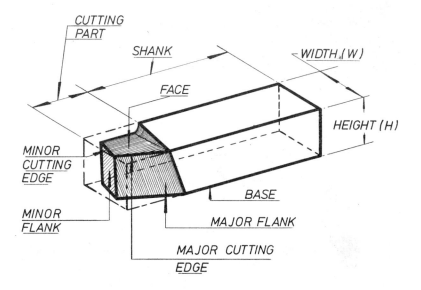

Fig. 3.1 Elements of a single-point cutting tool

Tool and working angles

BS 1296 gives two systems of reference:

a) A set of terms to enable the tool manufacturer to produce, sharpen, and measure the tool. This system is known as 'the tool-in-hand' system, fig. 3.2. With one exception (the normal wedge angle), all of the terms are identified by the use of the prefix 'tool' in the title, e.g. 'tool cutting-edge angle'.

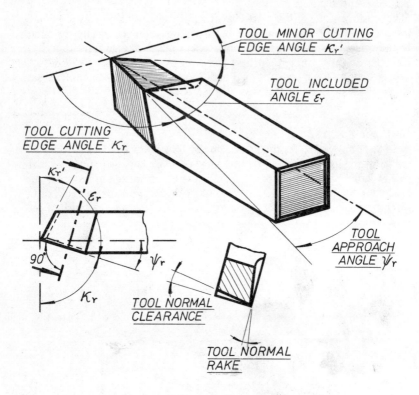

Fig. 3.2 The tool-in-hand reference system

b) A set of terms used to identify those angles that affect the tool when cutting, known as the 'tool-in-use' system. All of the terms in this system are prefixed by the word 'working' in the title, e.g. working normal rake, fig. 3.3.

BS 1296: Part 3: 1972 covers the size and shape of single-point cutting tools. Some of the more generally used shapes are shown in fig. 3.4. This standard recommends four tool normal rake angles: $0°$, $5°$, $10°$, and $14°$. Other rake angles can be obtained on standard tool shapes by arrangement with the cutting-tool manufacturer, or they may be ground on by the user as required.

34

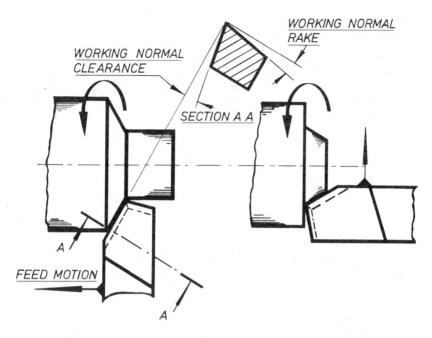

Fig. 3.3 The tool-in-use reference system

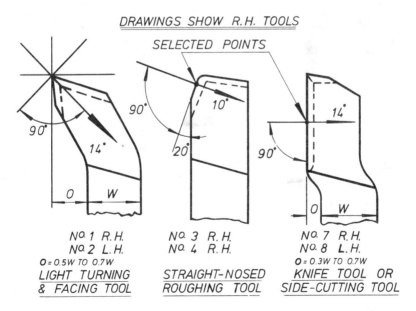

DRAWINGS SHOW R.H. TOOLS

SELECTED POINTS

NO. 1 R.H.	NO. 3 R.H.	NO. 7 R.H.
NO. 2 L.H.	NO. 4 R.H.	NO. 8 L.H.
O = 0.5W TO 0.7W		O = 0.3W TO 0.7W
LIGHT TURNING & FACING TOOL	STRAIGHT-NOSED ROUGHING TOOL	KNIFE TOOL OR SIDE-CUTTING TOOL

Fig. 3.4 Examples of standard tool shapes (*cont'd on page 36*)

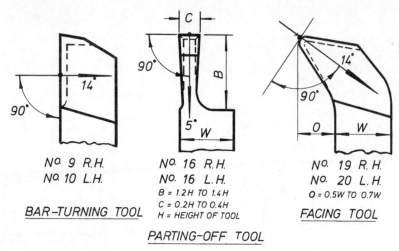

NO. 9 R.H.
NO. 10 L.H.

BAR–TURNING TOOL

NO. 16 R.H.
NO. 16 L.H.
B = 1.2H TO 1.4H
C = 0.2H TO 0.4H
H = HEIGHT OF TOOL

PARTING–OFF TOOL

NO. 19 R.H.
NO. 20 L.H.
O = 0.5W TO 0.7W

FACING TOOL

Fig. 3.4 Examples of standard tool shapes (*cont'd*)

3.2 Analysis of metal-cutting

For the purpose of simplifying the analysis of the metal-cutting process, it is convenient to consider it as being similar to the process shown in fig. 3.5. In this process, a cutting tool in the shape of a wedge is presented to the work material in such a way that relative movement between the work and the tool will cause the tool edge to be forced into the workpiece surface, resulting in part of the workpiece material being separated from the parent material. If the relative movement between the tool and work continues, then the metal removal (machining) will also be continuous.

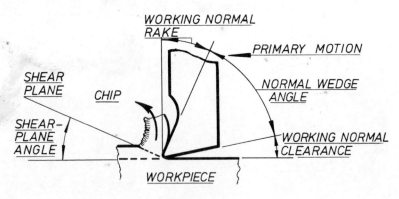

Fig. 3.5 Wedge-cutting action and shear plane

36

In the past it was thought that, as it was compressed, the material ahead of the tool became plastic (i.e. was irreversibly deformed, without breaking) and flowed along a line at approximately 90° to the tool face, this line being known as the *shear plane* (see fig. 3.5). Research has since shown that this concept is not strictly true, and it is now accepted that plastic deformation takes place in a zone around the shear plane, fig. 3.6. However, when dealing with the mathematical analysis of the cutting process it is more convenient to consider the shear-plane theory.

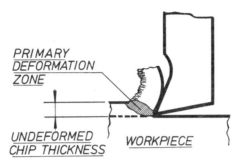

Fig. 3.6 Shear zone

Types of chip formation
The metal removed by the cutting process is called the *chip* and will vary in size, shape, and form according to the type of material being cut, the cutting-tool angle, and the position of the tool relative to the workpiece. The surface produced by the cutting process is called the *machined surface*, the original surface is called the *work surface*, and the intermediate surface produced due to the cutting process is known as the *transient surface*, fig. 3.7.

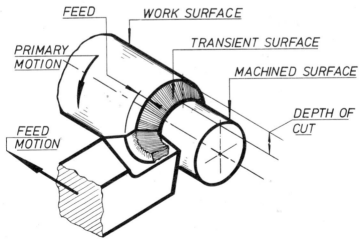

Fig. 3.7 General cutting terms

37

There are three different types of chip formation:
a) continuous chip,
b) continuous chip with a built-up edge,
c) discontinuous chip.

Continuous chip   This type of chip is formed when cutting ductile materials such as low-carbon steel, aluminium, and copper under good cutting conditions – using high speeds and fine feeds and with good lubrication. The cutting tool having a large working rake angle and polished faces will result in low friction between chip and tool and reduced power consumption, fig. 3.8.

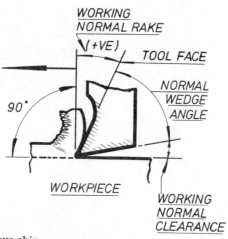

Fig. 3.8 Continuous chip

Continuous chip with a built-up edge   When cutting conditions are not correct, the friction between the chip and the tool face will be high, resulting in a rise in temperature causing the chip material to weld itself to the tool face. This in turn will give rise to further frictional resistance and a rapid build-up of chip material on the cutting edge of the tool. This accumulated build-up of chip material will then break away, part adhering to the underside of the chip and part to the workpiece. This process is repeated at a very high frequency, giving rise to a poor finish on the machined surface and accelerated wear on the tool face.

The condition usually occurs when cutting ductile materials, using a tool with a small working normal rake angle, with a poor surface finish on the tool face, and using a low speed with poor lubrication. This leads to cutting conditions of high temperature and pressure, fig. 3.9.

Discontinuous chip   When the work materials are brittle, or when cutting ductile materials at low speeds and high feeds, fracture occurs in the primary deformation zone (see fig. 3.6) and a discontinuous chip is formed. Cutting

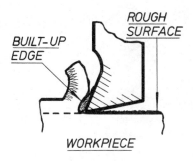

Fig. 3.9 Continuous chip with a built-up edge

tools produce this type of chip when using low cutting speed, large chip thicknesses (i.e. heavy feed rates) and small working normal rake angles (either positive or negative rake). The surface finish of the work is usually satisfactory, and the chips (swarf) are small enough to be easily handled and disposed of, fig. 3.10.

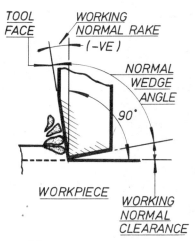

Fig. 3.10 Discontinuous chip

Orthogonal cutting
Orthogonal cutting occurs when the major cutting edge of the tool is presented to the workpiece perpendicular to the direction of the feed motion. The term 'orthogonal' was first used by Dr Merchant of the Cinncinatti Milling Machine Co. in a paper dealing with the 'Mechanics of the cutting process'. The attraction of orthogonal cutting is that only two forces are involved, which makes the analysis of the cutting motion much easier, fig. 3.11.

39

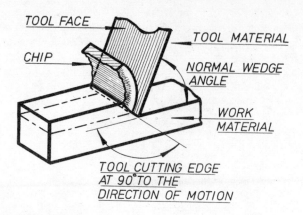

Fig. 3.11 Orthogonal cutting

Oblique cutting

This form of cutting occurs when the tool major cutting edge is presented to the workpiece at an angle which is not perpendicular to the direction of feed motion. This complicates the cutting analysis, fig. 3.12.

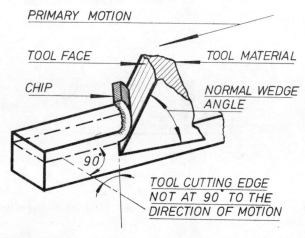

Fig. 3.12 Oblique cutting

3.3 Positive- and negative-rake cutting

One of the greatest limitations of the early cutting-tool materials was their inability to cut at high speeds. This was due to the fact that the heat generated by the action of the chip sliding over the tool face softened the tool material at fairly low temperatures (around 300 °C).

With modern cutting-tool materials such as ceramics and carbides (covered at the end of this chapter), although their hardness is vastly superior, even at high temperatures, they do have the inherent drawback of being brittle. This has led to what is known as *negative-rake cutting*, where the inclination of the tool face to the machined surface is greater than 90°. It can be seen from fig. 3.13 that this arrangement gives a much stronger and rigid tool, enabling it to withstand the considerably higher loads sustained when machining hard, strong, and tough materials at high speeds.

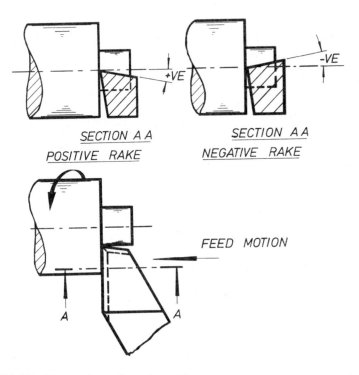

Fig. 3.13 Positive- and negative-rake cutting

When using negative-rake tipped tools, cutting speeds that remove chips in a red-hot condition are possible, as most of the heat generated by the cutting action is carried away by the chip and does not have time to pass on to the tool. The tool also has a greater bulk of material available to conduct away the remaining heat. Machining at high speed also promotes ideal chip formation, giving an excellent surface finish to the component being machined. Unfortunately all these factors make great demands on the machine tool carrying out the operation. Machining at high speed usually means removing a lot of metal, which requires a great deal of power, thus the machine must have a large motor. As mentioned above, ceramics and carbides tend to be brittle; hence the machine must be extremely rigid – any vibration would

41

very soon lead to tool failure. The machine must also have a suitably high range of speeds available. Many older machines are not really suitable to take full advantage of negative-rake cutting, due to the above limitations.

With positive-rake cutting tools, the tip of the tool is weaker and there is less material at the tip to conduct away the heat generated by the cutting action. The cutting speed has therefore to be limited so that an effective tool life (the length of time the tool is cutting) can be achieved.

It is perhaps relevant to mention here that, although the 'tool-in-hand' cutting angles may have the correct intended values for a given operation, when the tool is set in the machine the 'tool-in-use' angles may have widely different values according to the accuracy of set-up. The effect of setting errors above and below centres on the working normal rake and clearance angles can be seen by studying fig. 3.14.

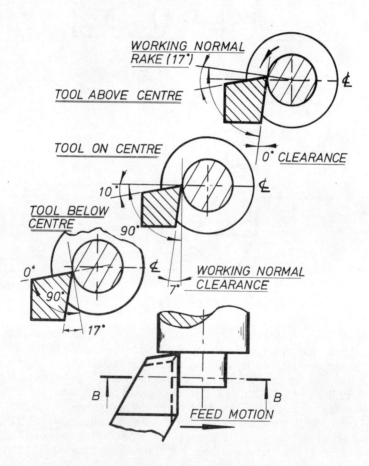

Fig. 3.14 Effect of tool setting on working normal rake and clearance angles

3.4 Cutting forces

In any material-cutting process, considerable forces are involved and, in order to design machine tools, tools, tool-holders, and work-holders that will withstand these forces, it is important to establish the magnitudes of the forces and the directions in which they act.

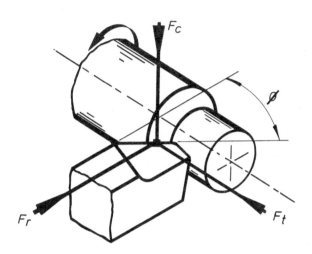

Fig. 3.15 Cutting forces, oblique

Figure 3.15 shows in diagrammatic form the forces acting at a selected point on the cutting edge of an oblique cutting tool. The vertical force component F_c is the force created by the primary motion, i.e. the motion of the work. Component F_t is the force created by the feed motion. Component F_r is the radial force. Each of these forces can be measured, but in order to simplify the measurement and analysis of the cutting test, angle ϕ is made $0°$ and F_r then becomes negligible. We now have a situation where the cutting forces acting are identical with those for an orthogonal cutting tool, fig. 3.16.

Here two forces are considered: the primary motion force F_c and the thrust force F_t. In practice, the thrust force is relatively small; in other words, most of the power is available for metal removal. By determining the force F_c, the power consumed during the cutting process can be calculated.

Measurement of cutting forces

The instrument used to measure cutting forces is called a *cutting-tool dynamometer*. There are several different designs of dynamometer, though their basic principle is the same. The cutting tool, supported in the instrument, is presented to the workpiece in the normal manner. As cutting takes place, the supporting elements deflect, i.e. they are subjected to strain, and, if these deflections or strains are measured accurately, the cutting forces in one, two, or three planes may be determined.

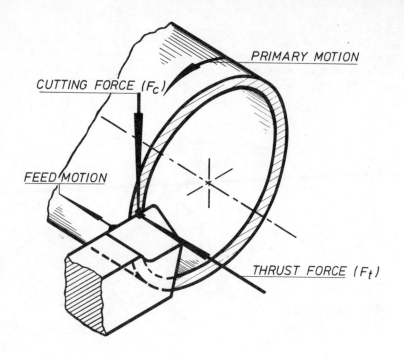

Fig. 3.16 Cutting forces, orthogonal

The design of a simply dynamometer is shown in fig. 3.17. With this type of dynamometer, the deflections caused by the cutting-tool forces are measured using dial-test indicators, the amount of deflection being proportional to the force acting upon the tool. Having established the forces acting on a cutting tool, this information can now be used to calculate the power consumed.

Power = rate of doing work

$$= \frac{\text{work done}}{\text{time taken}}$$

Work done = force x distance moved by force

Force = F_c = cutting force in newtons

Let mean cutting diameter = D mm

   speed of rotation of work = N rev/min

then distance moved by force = $\pi D N$ mm/min

$$= \frac{\pi D N}{1000 \times 60} \text{ m/s}$$

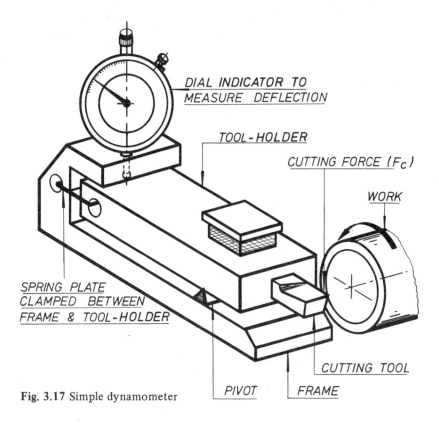

DIAL INDICATOR TO MEASURE DEFLECTION

TOOL-HOLDER

CUTTING FORCE (F_c)

WORK

SPRING PLATE CLAMPED BETWEEN FRAME & TOOL-HOLDER

CUTTING TOOL

PIVOT   FRAME

Fig. 3.17 Simple dynamometer

$\therefore$ work done per second = power

$$= \frac{F_c \times \pi D \times N}{1000 \times 60} \, \text{N m/s}$$

$$= \frac{F_c \times \pi D \times N}{1000 \times 60} \, \text{watts}$$

i.e. power consumed during cutting $= \dfrac{F_c \times \pi D \times N}{1000 \times 60}$ watts

Example Calculate the power consumed during the cutting of a low-carbon steel bar, 50 mm mean diameter, where the measured cutting force F_c is 2 kN at 100 revolutions per minute.

$$\text{Power consumed during cutting} = \frac{F_c \times \pi D \times N}{1000 \times 60}$$

$$= \frac{2000 \times \pi \times 50 \times 100}{1000 \times 60}$$

$$= 523 \text{ watts}$$

45

3.5 Factors affecting the cutting process

Experiments may be carried out varying such factors as cutting speed, depth of cut, feed rate, and cutting-tool geometry in turn, to obtain the most economical cost of production, taking into account such important factors as cutting-tool life, minimum power consumption, and lowest machining cost.

By carrying out cutting-tool tests using a dynamometer, the following may be easily verified. It will be found generally that when the tool normal rake angle is increased, the length of the shear plane is decreased and this in turn results in a lower cutting force F_c and therefore reduces the power consumed. Referring to fig. 3.18, it will be seen that as the tool normal rake increases, the tool normal wedge angle decreases and this will have the effect of giving a weaker tool section, an increased rate of tool wear, and consequently a rapid decrease in the tool life (i.e. the time the tool is actually cutting).

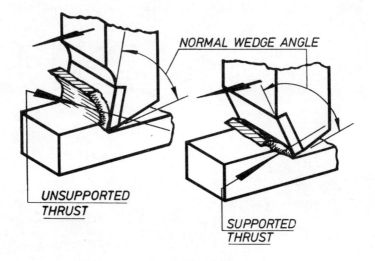

Fig. 3.18 Effects of altering tool normal wedge angle

Decreasing the tool normal rake angle will have the opposite effect to the above: the cutting force will increase and the power consumed during cutting will increase; the tool section will be stronger and the wear rate will decrease. This means that the conditions will be more suitable for the machining of high-tensile-strength materials such as stainless steel, nickel—chrome alloys, and so on.

If the cutting speed is kept constant, an increase in the area of cut will give an increase in the rate of metal removal. The cutting force F_c will increase, and more power will be consumed.

Decreasing the feed speed and/or the depth of cut reduces the power consumed, giving longer life to the tool cutting edge.

3.6 Cutting-tool materials

As stated earlier, metal cutting is one of the most important processes in industry. Having considered the factors that greatly affect the efficiency of the process, such as tool geometry, depth of cut, speed, and feed rates, it is now appropriate to consider another important factor in the cutting process — the cutting-tool material. The requirements of a cutting-tool material are

a) it must be considerably harder (resistant to wear, abrasion, and indentation) than the material being cut;
b) it must be strong enough to withstand the forces being applied due to cutting, i.e. bending, compression, shear, etc.;
c) it must be tough (resistant to shock loads) — this is very important when used for intermittent cutting;
d) it must be easily formed to the required cutting shape;
e) it must be capable of withstanding high temperatures;
f) it must be capable of withstanding the sudden cooling effect of coolant on the cutting tool.

The following materials will, in general, conform to the above requirements, and the material chosen for a particular application will depend on the material being machined, the quantity of components being produced, and the cost of the cutting-tool material:

a) carbon tool steel,
b) high-speed steel,
c) carbides,
d) ceramics,
e) diamonds.

Carbon tool steel

Plain-carbon steel containing 1 % carbon was one of the earlier forms of tool steel. It was forged to shape, hardened, and tempered to produce a cutting tool that would cut satisfactorily on non-ferrous materials and low-carbon steels. As the need for higher cutting speeds arose, and component materials became tougher and harder, this type of material was found to be unsatisfactory, since it started to lose its hardness at temperatures around 300 °C, even with the addition of alloying elements such as manganese, chromium, vanadium, etc.

Although carbon tool steel is used today for applications where high temperatures are not involved, it is generally being superseded by high-speed-steel tools, which, although they are more expensive initially, give improved cutting performance and higher metal-removal rates.

High-speed steel

High-speed steel (HSS) was introduced around 1900 by F. W. Taylor of the Bethlehem Steel Corporation, who developed a heat-treatment process to produce this type of steel. High-speed steels are alloy steels containing tungsten (18 %), chromium (4 %), vanadium (1 %), carbon (0.7 %) and other elements

47

such as molydenum and cobalt. This type of cutting-tool material will give
excellent performance over a great range of materials and cutting speeds, and
it retains its hardness up to around 600 °C. Furthermore, by modifying the
proportions of the alloying elements, cutting tools may be produced for
specific cutting applications; for example, by increasing the tungsten content
to 20 % or higher, a super-grade HSS will be produced which will cut at high
temperatures and have better wear resistance than the ordinary grade of HSS.

Cemented carbides

These cutting tools consist of tungsten, tantalum, columbium, and titanium
carbides, together with a 'binder', usually cobalt, all mixed together as fine
powders. These powders are compacted (compressed) into the required shape
and are subjected to a high-temperature treatment known as 'sintering'. During
this process, the cobalt binder is fused to the carbides, producing a hard dense
substance that is the basis of the cutting tool. Cemented carbides, are very
hard, and the usual practice is to confine the size to a relatively small shape,
known as an 'insert', which is clamped to a tough steel shank or holder,
fig. 3.19. This has the advantage that the tool bit is well supported to resist
the cutting forces. The insert is designed so that each of its cutting edges can
be used in turn. It may then be discarded and replaced with a new tool bit,
giving low maintenance and breakdown time.

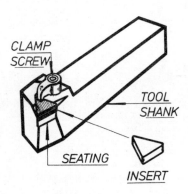

Fig. 3.19 An inserted carbide-tipped tool

Ceramics

Ceramics provide a further example of how modern materials development is
enabling the tooling engineer to obtain even higher machining rates — machining rates with ceramics are two or three times faster than those with cemented-
carbide tools, and cutting speeds on cast iron in the order of 1000 m/min are
not uncommon.

Aluminium oxide and boron nitride powders mixed together and sintered
at 1700 °C form the ingredients of ceramic tools. These tools are extremely

hard, with good compressive strength, but unfortunately they are also very brittle. This means that great care must be exercised when considering ceramics for a machining operation, particularly where vibrations, shock loading, and intermittent cuts are involved, for under these conditions the tool is certain to fail. However, for straightforward cutting where rigid machines with sufficient power and adequate work-holding facilities are available, ceramic tools will operate at very high speeds and will produce an excellent surface finish.

Diamonds

Industrial diamonds are among the hardest materials known. They have a low coefficient of friction, high compressive strength, and are extremely wear-resistant. Used mainly for cutting non-ferrous materials, diamond tools will give very good surface finish at high speeds, with good dimensional accuracy. They are produced by clamping or bonding (with epoxy-resin adhesives) small chips of diamond to the tool shank and lapping to the required shape.

Diamond tools tend to be small and best suited to light cuts and finishing operations where there is little shock loading. Unfortunately, these tools are not easily serviced in the average tool room, and when a diamond tool shows signs of wear it is best sent back to the manufacturer, who will have the necessary equipment and expertise to relap it.

Summing up, it is fair to say that it is the cemented-carbide and ceramic tools that must be looked to for increased production. There is indisputable evidence that the application of these tools is increasing rapidly over the whole range of material-removal processes, and this is likely to accelerate considerably in the near future. However, this statement must immediately be qualified by emphasising that the prime requirements when applying these types of tooling are (a) adequate machine-tool power and (b) substantial and robust work- and tool-holding facilities. Conditions not meeting these requirements will result in the tools not being used to their full capability, and expensive tool failure.

3.7 Functions of a cutting fluid

When metal is being removed from a component by a process using a single-point cutting tool, heat and wear are inevitably produced, caused by friction as the chip passes over the tool face. Heat is also produced by the internal shearing action that takes place as the chip is being formed (known as plastic flow of the material) and by the action of the workpiece in contact with the cutting tool, fig. 3.20.

Heat and wear are undesirable characteristics of the metal-cutting process — eventually causing failure of the tool — and the amount generated will vary according to the conditions operating at the time of cutting. In order to obtain a reasonable tool life (i.e. the length of time that the cutting tool will cut satisfactorily), cutting speeds have to be regulated to keep down the temperature of the cutting tool and/or workpiece. If no limitation were

49

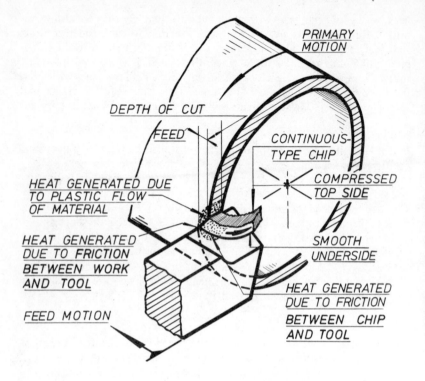

PRIMARY MOTION

DEPTH OF CUT

FEED

CONTINUOUS-TYPE CHIP

COMPRESSED TOP SIDE

HEAT GENERATED DUE TO PLASTIC FLOW OF MATERIAL

HEAT GENERATED DUE TO FRICTION BETWEEN WORK AND TOOL

SMOOTH UNDERSIDE

FEED MOTION

HEAT GENERATED DUE TO FRICTION BETWEEN CHIP AND TOOL

Fig. 3.20 Tool and chip, showing areas of heat generation

placed on the cutting speed then either the tool would overheat and soften and the cutting edge would break down or, in some cases, the work material would suffer damage — for example, in the case of plastics and some zinc-based alloys, the surface would start to melt; in the case of some magnesium alloys, there would be a danger of the workpiece and the swarf catching fire.

One way of improving the metal-cutting operation is by using a cutting fluid. A cutting fluid has two primary functions: *cooling* and *lubricating*. The cooling function is achieved by supplying an adequate volume of cutting fluid to carry the heat away from the cutting tool and workpiece. By lubricating the cutting tool, the coefficient of friction between the chip and the cutting tool is reduced, effectively reducing the temperature and the amount of mechanical wear that takes place, making the cutting process more economical.

There are also several secondary benefits from using a cutting fluid, their value depending upon which type of fluid is chosen for a particular application (the different types of fluid will be dealt with in section 3.8). Among these secondary benefits are

a) the use of a cutting fluid invariably results in a better surface finish;
b) the tool life is increased;

c) the cutting fluid removes heat, thereby reducing workpiece expansion, making possible more accurate production and measurement with less distortion;
d) where cutting conditions are suitable for the formation of a built-up edge, an appropriate cutting fluid will reduce or eliminate this effect;
e) power consumption is reduced, due to the introduction of a lubricant between chip and tool;
f) the cutting fluid washes away swarf, giving easier swarf control and, in the case of cast iron, dust-free control;
g) modern cutting fluids prevent rusting or staining of both the workpiece and the machine tool, though certain fluids (those containing sulphur) must be avoided when machining cuprous materials, i.e. materials containing copper.

3.8 Types of cutting fluid

Most machining operations can advantageously be carried out using a cutting fluid, for the reasons given above. However, the choice of a suitable fluid must be carefully considered, as each of the many types has been formulated for a particular service condition.

Generally, cutting fluids are divided into two main types:

a) water-based fluids;
b) straight or neat oils (this is a confusing term as many of the oils are not used neat but are mixtures of oils or oils with chemicals such as sulphur and chlorine added to them).

Water-based cutting fluids

These are the most widely used types of cutting fluid, the most popular being 'soluble oil' (suds or slurry).

Soluble oil (1 % to 5 %) is mixed with water to form an emulsion. This has excellent cooling properties at low cost, and there is also some lubricating effect between tool and chip, thus tool wear is reduced.

Modern soluble oils contain a corrosion-inhibitor and a biocide to keep down the growth of bacteria that would otherwise become a health hazard. In addition, additives can be included to enable the fluid to perform satisfactorily under more arduous conditions.

Less frequently used forms of water-based cutting fluids are based on chemical solutions. Soda solutions are often used on grinding operations — they have a good flushing action and cool well.

Straight or neat oils

These have several subgroups:

a) neat mineral oils,
b) mixtures of mineral and fatty oils,
c) sulphurised oils,
d) oils with sulphur and/or chlorine additives.

51

The selection of a particular type is not easy. Many factors have to be taken into account, such as the depth of cut, the cutting speed, the feed rate, the tool material, the work material, and the tool life required (this may be a more important factor than the maximum cutting speed). One large oil company lists over forty different types of cutting fluid in its literature, each one no doubt developed for a specific purpose.

The following observations may be used as a guide when selecting the correct type of cutting fluid.

a) Mineral oils   These are suitable for light machining operations where the cutting pressures are low. Typical applications would be for capstan and turret lathes and single-spindle automatics where free-cutting brasses and steels are being machined.

b) Mineral and fatty oil mixtures   Due to the addition of the fatty oils, these have higher film strengths (i.e. they are less easily squeezed out of the tool—work area) and they are suitable for heavier-duty operations, such as threading on capstan and turret lathes, thread milling, and medium-capacity automatic lathes.

c) Sulphurised oils   These are used for heavy-duty lathe work and gear-cutting, thread-grinding, and screw-cutting operations.

d) Sulphur and chlorine additives   These additives give the oil an E.P. (extreme pressure) property and are suitable for severe cutting operations on strong and tough materials, such as stainless steels and nickel alloys. They are also successfully used on broaching operations.

Although we have seen that advantages can be gained by using a carefully chosen cutting fluid, it must be realised that a cutting fluid costs money, even a water-based one. Care must be taken to ensure that it remains clean and does not become contaminated by sludge or deposited rubbish. Coolant tanks should be regularly emptied and thoroughly cleaned, and if this is done then the fluid will continue to be effective and will not become a health hazard.

3.9 Application of cutting fluids
When used, a cutting fluid must also be correctly applied. It is most important that a copious supply is maintained to the correct place, i.e. to where the heat and wear are being created — at the chip—tool interface. An intermittent supply caused by a blocked supply system or a low reservoir level can do much harm: when the supply stops, the temperature of the tool quickly rises and when the supply returns it has a quenching effect on the tool which may lead to cracks on the cutting edge.

Most of the larger modern machine tools have an adequate coolant system, usually enclosed within the body of the machine itself. The fluid is well filtered and is stored in a light-proof and cool area, from where it is pumped through a piping system to the area of tool—work contact. With care and regular maintenance, this arrangement gives trouble-free operation.

There are, however, machining operations that take place where a built-in coolant system is not available, and in these circumstances other methods of applying cutting fluids are required. The commonest method is to apply the cutting fluid from a small can to the work area by means of a brush. Discarded washing-up-liquid squeeze bottles and oil cans are also widely used, and aerosol-spray cans containing a cutting lubricant are also available. It must be pointed out most strongly that, as these methods, involve the introduction of a third element into the work—tool cutting area, they are potentially *dangerous*, and great care must be taken when applying the cutting fluid. If at all possible, the spindle should be stopped before the fluid is applied. If this is not feasible then it is preferable to direct a jet of cutting fluid to the cutting area. This is a safer method than using a brush, and greater quantities of cutting fluid can be directed to the cutting tool.

Exercises on chapter 3
1   Sketch a single-point lathe tool, naming the relevant parts and angles in accordance with the 'tool-in-hand' reference system.
2   State the three recognised types of chip formation. Sketch each type of chip and describe the conditions of cutting that lead to the formation of each type.
3   Explain the difference between orthogonal and oblique cutting.
4   Sketch an oblique single-point lathe cutting tool and indicate the forces acting on the tool during cutting.
5   Calculate the power consumed during cutting a 75 mm diameter bar at 130 revolutions per minute, using a single-point lathe tool cutting orthogonally — depth of cut 5 mm and measured cutting force (F_c) 3 kN.
6   List the main requirements for a cutting-tool material.
7   State the primary and secondary functions of a cutting fluid.
8   Show with the aid of a simple sketch the areas of heat generation that occur when using a single-point cutting tool on a lathe.
9   List the reasons why an engineering manufacturer should use a cutting fluid.
10   What is meant by the term 'water-based cutting fluid'.
11   Straight or neat oils are subdivided into several different types. Name four types and specify an application for each type.
12   Using a diagrammatic sketch or a block diagram, show a typical 'built-in' coolant system for a lathe or milling machine. Indicate such elements as the pump, reservoir, pipework, tap, the return flow, and the on/off electrical connections.
13   Define a multi-tooth cutter and name three examples of multi-tooth cutters.
14   Explain briefly the importance and procedures of mixing and storing cutting fluids.
15   Describe the relative merits of solid and 'throw-away-tip' cutting tools.

4 Machine tools

4.1 Elements of a machine tool

Part of the definition of a metalworking machine tool given in BS 4640: 1970, 'Classification of metalworking machine tools by type', is that 'a metalworking machine tool is a power-driven machine not portable by hand while in operation which works metal by cutting or forming'. This British Standard groups metalworking machine tools by type, for instance 'metal-cutting machine tools' and 'metal-forming machine tools'. In this chapter we are concerned only with metal-cutting machine tools.

The main function of a metal-cutting machine tool is to remove unwanted material and to control the geometry and the dimensional accuracy of the finished workpiece. This is achieved by moving the tool and the workpiece in the correct relationship to each other.

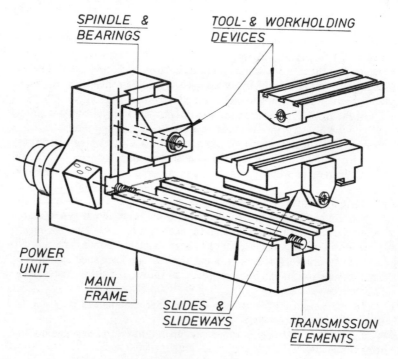

Fig. 4.1 Basic elements of a machine tool

A machine tool is fundamentally an assembly of basic elements, fig. 4.1. Machines of different types can be designed to perform specialised functions such as drilling, turning, milling, etc., but, nevertheless, the elements described below are usually essential features of each type of machine.

1 The main frame or structure

This element (often referred to as the 'bed' or 'body') acts as a base for the rest of the units. Beds or bodies are usually made from cast iron, which is a relatively cheap material and ideal as far as machines are concerned since it has a natural tendency to damp out any vibrations set up by the cutting or out-of-balance forces of the moving parts. The bed invariably has one or more slideways cast as an integral part of it, and for this cast iron is an advantage as it is easily machined, has ideal wearing properties, and, if required, a hard surface can be produced by a process of induction hardening (not covered in this book).

2 Slides and slideways

A slide is a moving element providing a straight-line movement to a workpiece or a tool-holder at a controlled feed rate. A slideway gives accurate guiding constraints to the slide and may contain some form of adjustment to eliminate play between the two members as wear takes place. The actual shape of the slideway will vary according to the type of machine, the loads being carried, and the degree of accuracy required. The more common types of slideway are shown in fig. 4.2; they include the 'vee', 'dovetail', 'flat', 'roller', and combinations of these.

Figure 4.2(a) shows a combination of vee and flat slides which provides the necessary degrees of constraint, is simple to manufacture, and is easily rescraped when wear has taken place. Variations of this basic form are used for the saddle and tailstock slideways on a lathe.

The dovetail slide, fig. 4.2(b), is quite often used where the cutting forces would tend to separate the two elements of a slide, causing vibration. With dovetail slides, the gib strip can be adjusted so that the two elements of the slide are a good sliding fit. It is the type of slide used on a centre-lathe cross-slide and compound slide.

Flat slides, fig. 4.2(c), have a larger load-carrying capacity suitable for the larger type of machine. They provide large areas of contact, thus wear is kept to a minimum.

Roller slides, shown in fig. 4.2(d), reduce the friction between the two members of the slide, making it easy to move heavy loads positively and accurately.

There are also types of slideway design, including hydrostatic and pneumatic types, where the two elements of the slide are separated by a fluid film.

3 Spindles and spindle bearings

These control the rotational movement of cutting tools and/or work-holding devices, the accuracy of this movement being dependent upon the degree of precision of the spindle and spindle bearing.

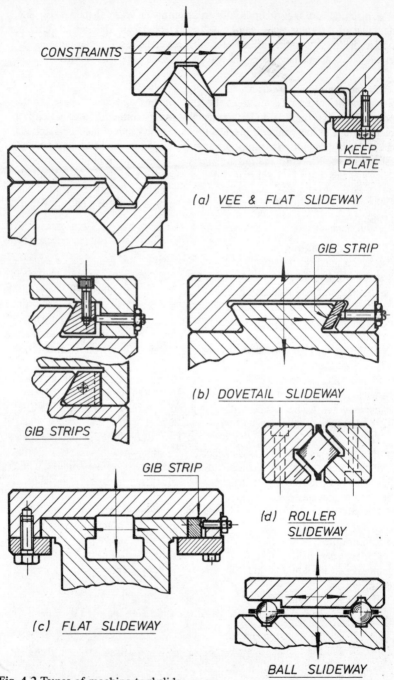

CONSTRAINTS

KEEP PLATE

(a) VEE & FLAT SLIDEWAY

GIB STRIP

GIB STRIPS

(b) DOVETAIL SLIDEWAY

GIB STRIP

(d) ROLLER SLIDEWAY

(c) FLAT SLIDEWAY

BALL SLIDEWAY

Fig. 4.2 Types of machine-tool slide

4 Power units

The power unit of the modern machine tool is the electric motor. There may be only one motor from which other movements are transmitted, or there may be a main motor with separate drive motors for individual elements, such as a coolant-pump motor or a feed-movement motor.

5 Transmission elements

These transmit and control the essential movement from the motor (or from the operator in the case of manual operation). They include the drives to the spindle and the feed drives. These elements may be mechanical (pulleys, belts, screws and nuts, etc.), hydraulic, or pneumatic.

Slideways are usually controlled by a leadscrew-and-nut arrangement which may be hand- or power-driven. Larger leadscrews have a square or acme-type thread, although smaller ones sometimes use a vee thread. Two typical arrangements are shown in fig. 4.3. At (a), the lathe leadscrew has only one degree of movement, that of rotation; the half nut also has only one movement — linear — moving the saddle along the bed of the lathe. With the milling-machine slideway system shown in fig. 4.3(b), the nut is stationary and the leadscrew not only rotates but also has a linear movement.

Although the leadscrew and nut is a relatively simple way of obtaining a slide movement, it is sometimes necessary to control the amount of displacement that takes place, e.g. to control the depth of cut on a lathe cross-slide. A common way of achieving this is to fit a calibrated dial on the end of the leadscrew, fig. 4.4. If a leadscrew with a 5 mm pitch is fitted with a dial divided into 100 equal parts, then one division will represent $\frac{5}{100} = 0.05$ mm. When using this type of arrangement for controlling dimensional accuracy on a machine tool, it is essential to realise that, to function satisfactorily, a leadscrew-and-nut system has a certain amount of clearance. It is therefore necessary to use the dial in one direction only, to avoid errors caused by backlash.

The standards of accuracy expected from slideway systems (and from metal-cutting machine tools in general) are covered by BS 3800: 1964, 'Methods for testing the accuracy of machine tools', and BS 4656, 'The accuracy of machine tools and methods of test', which is published in several parts covering a range of machines. In general, the accuracy is tested by a number of geometrical checks using relatively straightforward equipment such as dial-test indicators, test mandrels (test bars), straight-edges, spirit-levels, etc. to establish that a particular element of a machine tool is within the tolerances specified in BS 4656.

It must be appreciated that the accuracy of a slideway system does not depend upon the type of system (i.e. vee, dovetail, flat, etc.) but upon how well the system is designed and produced. Also, the accuracy of a component produced on a particular machine will not necessarily be comparable with the accuracy of a single machine-tool element such as a slideway — other

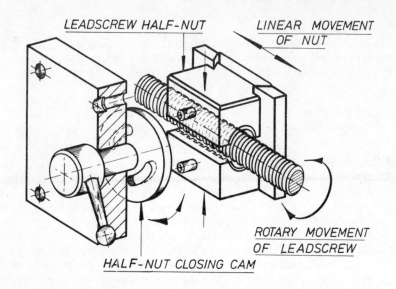

Fig. 4.3(a) Leadscrew-and-nut drives (moving nut)

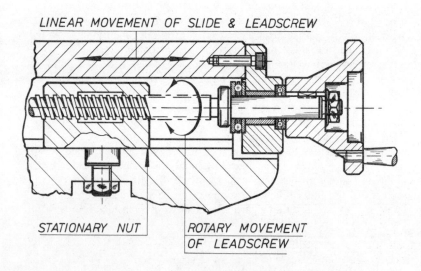

Fig. 4.3(b) Leadscrew-and-nut drives (moving screw)

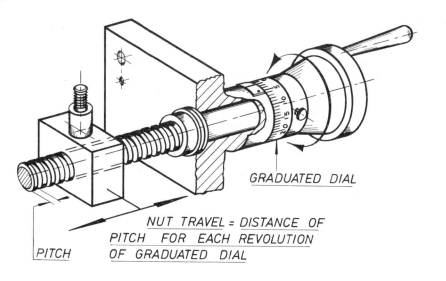

GRADUATED DIAL

NUT TRAVEL = DISTANCE OF
PITCH FOR EACH REVOLUTION
OF GRADUATED DIAL

PITCH

Fig. 4.4 Leadscrew calibration

factors such as the amount of heat generated, the amount of tool wear, the
inherent rigidity of the machine tool, and the accuracy of the device position-
ing the slide relative to the workpiece all ultimately affect the accuracy of
the finished workpiece.

6 Work-holding and tool-holding elements

These essential items on any machine tool are too numerous to itemise in
detail. They may be of a standard nature, such as drilling- and shaping-machine
vices, or three- and four-jaw chucks on a lathe, or just tee slots on a machine
table from which a skilled craftsman can derive an individual work-holding
set-up for a particular job.

In the case of tool-holders, these are usually designed to suit a particular
type of tool, such as a morse-taper socket on a drilling-machine spindle, to
suit taper-shank drills and reamers etc. Some of these devices will be dealt
with when dealing with individual machine tools.

4.2 Degrees of freedom

The production of a workpiece on a machine tool involves holding the work
material and the tool (or tools) in a predetermined manner and controlling
the relative movement of these elements to produce a dimensionally accurate
geometric shape. Before considering these movements with respect to machine
tools, it is perhaps helpful to consider the principle of the six degrees of
freedom of a body in space. This principle states that a body in space has six
degrees (or paths) of pure movement: three degrees of linear movement,
conventionally designated 'X', 'Y', and 'Z'; and three degrees of rotational

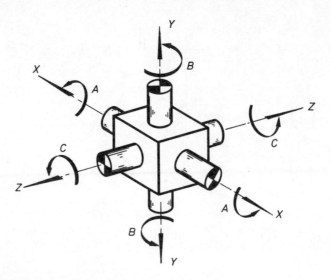

Fig. 4.5 Six degrees of freedom of a body in space

movement, designated '*A*', '*B*', '*C*' – fig. 4.5. All other movements are a combination of these degrees of freedom.

All movements on a machine tool are made up from one or more of these six degrees of freedom. The '*Z*' axis of motion on a machine tool is usually parallel to the axis of the main spindle, and the '*X*' axis is if possible arranged horizontally and parallel to the work-holding surface. The '*Y*' axis obviously has to be mutually perpendicular to these other two. Figure 4.6(a) shows this convention applied to a centre lathe, and fig. 4.6(b) to a vertical milling machine.

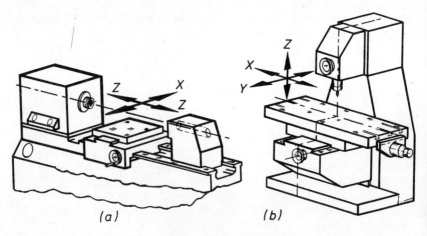

Fig. 4.6 Machine-tool axes: (a) centre lathe, (b) vertical milling machine

4.3 Generating, forming, and copying

The components produced by machine tools are essentially combinations of mainly simple geometric shapes such as the cylinder, cone, sphere (or part of a sphere), flat surfaces, and helical surfaces. In the previous section it was stated that machine-tool movements are made up from one or more degrees of freedom: by carefully controlling these movements relative to each other, it is possible to produce the geometric shape desired. An example of this is the production of a cylindrical shape on a lathe, where the work is rotated and the tool movement is controlled so that the tool travels in a path parallel to the axis of the work, producing a cylindrical shape, fig. 4.7. Such a process, where the shape of the work is controlled by a combination of work and tool movements, is called *generating*. Another example of generating is shown in fig. 4.15, where the flat surface produced by the shaper is a result of the straight-line movement of the cutting tool and the feed movement of the work. In both of these examples, the shape produced is independent of the shape of the tool.

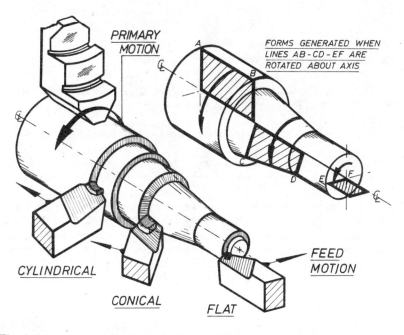

Fig. 4.7 Generating cylindrical, conical, and flat shapes

If the shape of the work does depend upon the shape of the tool, then a process known as *forming* is involved. Extrusion (fig. 1.9) is an example of true forming, where the shape of the work produced is a direct replica of the extrusion die.

In most machining operations involving forming, some generating is also inevitable; for example, a screw thread requires a form tool to produce the shape of the thread but requires a precise linear movement of the tool (the pitch distance) for each revolution of the workpiece to produce the helix which, by the previous definition, is a 'generated' shape, fig. 4.8.

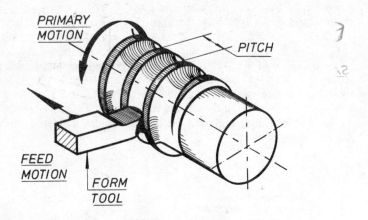

Fig. 4.8 Example of forming and generating combined

Figure 4.24 also illustrates the process of forming, the shape produced on the component in this case being dependent upon the shape of the form milling cutter.

There is one further system, called *copying*, which is used for producing more complicated shapes, the shape of the work being a replica of a 'master component' or accurately made model or template. Copying is really a special case of generating, requiring some form of copying system which can accurately control the tool movements relative to the work movement to produce a workpiece identical to the 'master' within the limits of accuracy of the copying system. Such systems are used mainly on lathes and milling machines and use a hydraulic, pneumatic, or electrical servomechanism to control the tool movements. (A servomechanism is a control system requiring a relative small input signal to provide a larger output force to operate the mechanism.)

4.4 Capstan and turret lathes
Machine tools have evolved over the last 200 years or so, new machines being developed as particular production requirements arise. The earliest lathes were very simple, slow, and not very accurate, and the production of components was usually on a 'one-off' basis.

Capstan and turret lathes were developed from the centre lathe, mainly in America, as larger quantities of similar components to a higher degree of accuracy were demanded. The essential difference between the centre lathe

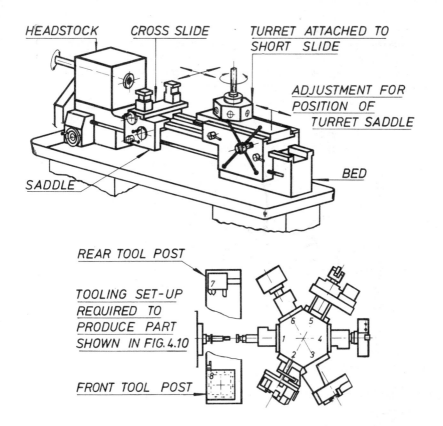

Fig. 4.9 Block diagram of a capstan lathe

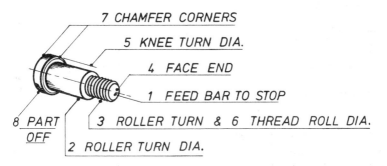

Fig. 4.10 Typical component made on a capstan lathe

and the capstan and turret lathes is that the tailstock of the centre lathe is replaced by a multi-tool turret providing a number of tooling stations which can be presented to the workpiece in rapid succession. Figure 4.9 shows the eleven tooling positions available on capstan and turret lathes. In some of these positions, tool-holders with more than one cutting tool may be used.

The cutting tools required for a particular component are preset to stops by a skilled setter, enabling the machine to be operated by an unskilled worker, producing components quickly and relatively cheaply. A typical example is shown in fig. 4.10.

Turret lathes tend to be larger than capstan lathes, the main design difference being that with turret lathes the turret saddle bears directly on the bed of the machine; thus the head (and the tools) can traverse the full length of the bed if required, fig. 4.11. With capstan lathes the turret head is mounted on a short slide with only a limited amount of travel.

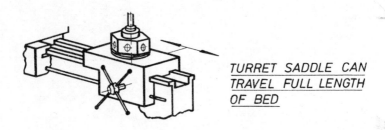

TURRET SADDLE CAN
TRAVEL FULL LENGTH
OF BED

Fig. 4.11 Saddle and slide arrangement on a turret lathe

A wide range of operations is possible with this type of machine, including turning, facing, centre-drilling, drilling, reaming, threading, boring, knurling, and parting-off.

4.5 The drilling machine

Drilling machines are among the simpler forms of machine-tool, their main function being to originate a hole, usually using a two-flute twist drill. Other tools such as counterbores, countersinks, reamers, and spotfaces are used in conjunction with the drilling machine to modify the original hole for a particular purpose, fig. 4.12. Figure 4.13 shows a block diagram of a drilling machine.

Drilling machines are provided with a range of speeds to suit the sizes of drills the machine will accommodate. With many machines, the force to feed the drill into the work is supplied by the operator — this type of machine is known as a 'sensitive' drilling machine, as the operator can feel, or sense, the progress of the drill throughout the operation. Other machines are equipped with a powered feed mechanism as well as the hand-feed arrangement.

In use, drilling machines are used either by a skilled craftsman on 'one-off'-type jobs to previously marked out positions, or on production work using some form of work-holding and drill-guiding device known as a 'drill jig'.

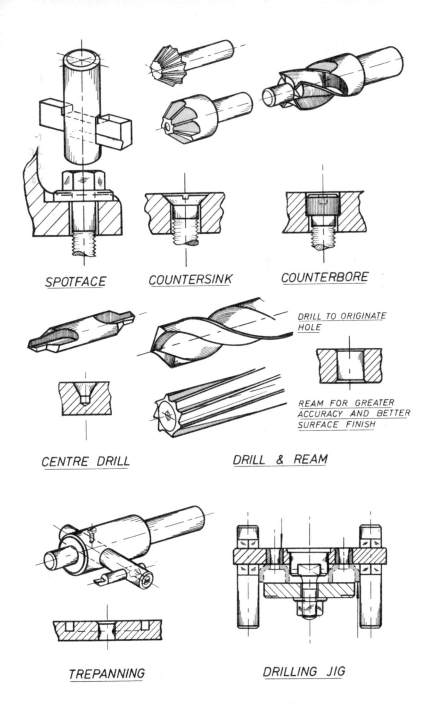

SPOTFACE   COUNTERSINK   COUNTERBORE

DRILL TO ORIGINATE HOLE

REAM FOR GREATER ACCURACY AND BETTER SURFACE FINISH

CENTRE DRILL   DRILL & REAM

TREPANNING   DRILLING JIG

Fig. 4.12 Drilling-machine operations and tools

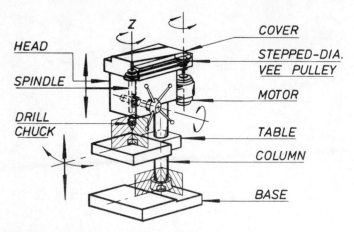

Fig. 4.13 Block diagram of a drilling machine

4.6 The shaping machine
The main function of a shaping machine, fig. 4.14, is the removal of metal by the use of a single-point cutting tool to generate a flat surface, fig. 4.15.

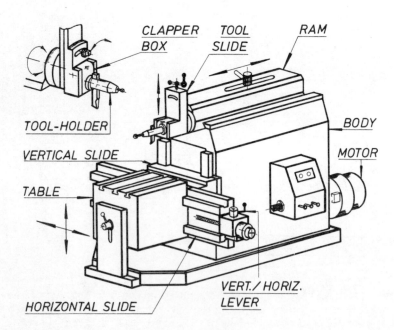

Fig. 4.14 Block diagram of a shaping machine

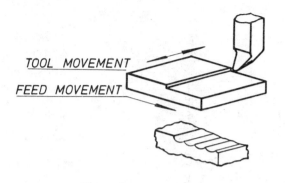

Fig. 4.15 Generation of flat surfaces on a shaping machine

A shaping machine in a well maintained condition, set up and operated by a skilled craftsman, can also be used to produce parallel surfaces and angular and square faces. Grooves or other formed shapes may be produced by using a tool having the correct form ground on it, but this type of operation is not very often required.

The component shown in fig. 4.16 is a typical example of work that can be produced on a shaping machine.

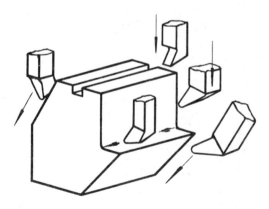

Fig. 4.16 Typical component made on a shaper

4.7 The grinding machine

This is a metal-cutting machine tool. Its prime function is to obtain dimensional and geometric accuracy and to improve the surface finish of the component being machined. It is often used during the final stages of the production process to finish to size components that have been shaped by other means

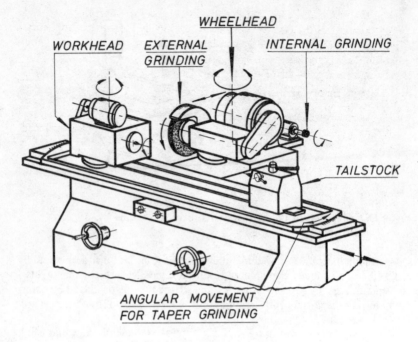

Fig. 4.17 Block diagram of a cylindrical grinding machine

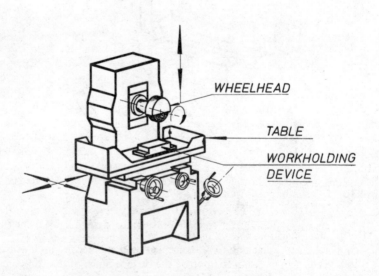

Fig. 4.18 Block diagram of a surface grinding machine

such as shaping, turning, milling, etc. It has the great advantage of being able to machine metals that are too hard to be machined by other methods (often after heat treatment for hardness or when a material is naturally hard, like tungsten carbide for instance).

Apart from the 'off-hand' grinder used to sharpen cutting tools by hand, there are several different types of grinder designed to machine specific geometric shapes. The commonest of these are the cylindrical grinder, fig. 4.17, and the surface grinder, fig. 4.18.

A typical example of grinding work would be the finishing of a hardened steel drill bush, fig. 4.19.

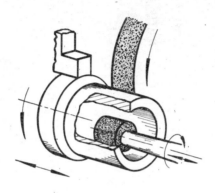

Fig. 4.19 Typical component finished on a grinding machine

4.8 The milling machine

The basic milling machines illustrated in fig. 4.20 show that the milling machine is primarily capable of linear movement along the X, Y, and Z axes. It is possible, however, to obtain other than straight-line cuts by the use of accessories such as the dividing head and the rotary table, fig. 4.21.

There are several different types of milling machine, but this book will be concerned only with the vertical and horizontal types. The milling cutters, which may be mounted on an arbor or directly in the spindle nose (depending upon the type), are rotated to provide the necessary cutting action. The work must be securely fixed to the table, either directly or indirectly with some form of work-holding device such as a machine vice, rotary table, dividing head, or special fixture.

Geometrical shapes can be generated by moving the work past the revolving cutter, and these include flat and helical surfaces, fig. 4.22. By combining a series of flat faces using the dividing head, geometrical shapes such as hexagons, squares, etc. may easily be obtained, fig. 4.23. Formed shapes are obtained by using specially shaped cutters, and typical examples are shown in fig. 4.24.

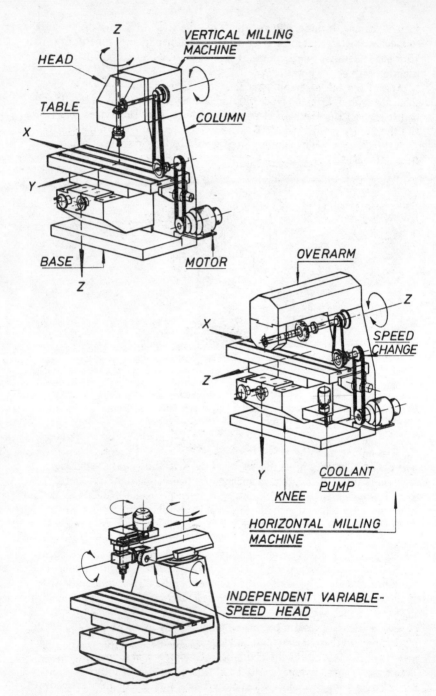

Fig. 4.20 Block diagrams of vertical and horizontal milling machines

70

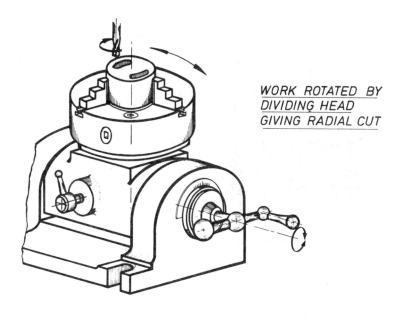

WORK ROTATED BY
DIVIDING HEAD
GIVING RADIAL CUT

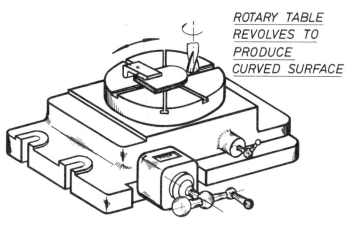

ROTARY TABLE
REVOLVES TO
PRODUCE
CURVED SURFACE

Fig. 4.21 Dividing head and rotary table

Exercises on chapter 4
1   State the purpose and functions of a machine tool.
2   Give three reasons why cast iron is the material mainly used for the production of machine beds.
3   Make neat sketches showing three different forms of slide and slideway design. Where appropriate, indicate methods of adjusting for wear.

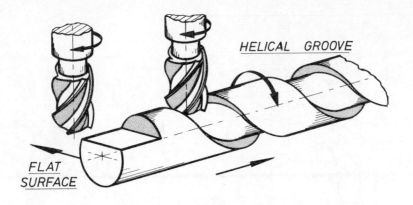

FLAT
SURFACE

HELICAL GROOVE

Fig. 4.22 Generating surfaces on the milling machine

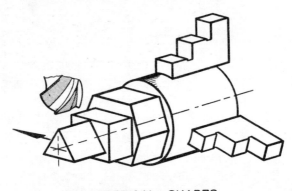

GEOMETRICAL SHAPES

Fig. 4.23 Production of geometric shapes on the milling machine

4 With the aid of a diagrammatic sketch, show the line of transmission from the motor to the chuck and cross-slide on a centre-lathe.

5 A lathe cross-slide leadscrew has a pitch of 2 mm. How many graduations would be required on the leadscrew dial if each graduation represented 0.01 mm movement of the cross-slide?

6 Illustrate how a dovetail slideway system eliminates all degrees of freedom except one linear movement required for slide movement.

7 Why can a capstan lathe, once it has been set-up, produce components at a faster rate than a centre lathe?

8 State the range of machining operations that can be carried out on a sensitive drilling machine.

9 With the aid of a sketch, show the line of transmission from the motor to the ram and feed-operating mechanism on a shaping machine.

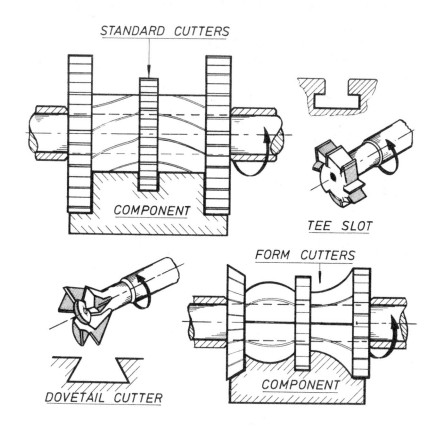

STANDARD CUTTERS

COMPONENT

TEE SLOT

FORM CUTTERS

DOVETAIL CUTTER

COMPONENT

Fig. 4.24 Production of formed shapes on the milling machine

10   State the main advantages of grinding machines over other types of metal-machining machine tools.

11   Sketch three different types of work-holding methods used on a milling machine.

12   List the machining sequence of operations required to produce a small pair of toolmakers' clamps.

13   Explain the main difference between producing a generated shape and a formed shape on a milling machine.

14   State the main function of a metal-cutting machine tool.

15   Explain the essential difference between a rotary table and a dividing head.

5   Measurement

5.1 The purpose of measurement

The basic purpose of measurement in the engineering industry is to determine whether a component has been manufactured to the requirements of a specification, usually an engineering drawing. There are many types of measurement which may be made, but this chapter will be mainly concerned with the measurement of length and angle, and with certain geometrical qualities (form) such as roundness, squareness, and flatness.

Measurement is essentially a process of comparison between the component being measured and a known standard. If by this process the component is found to be outside the drawing specification, then it will need to be scrapped or reworked, depending on which features are found to be incorrect. The accuracy of the known standard is therefore obviously of special importance, and the choice of a measuring method will depend on the accuracy demanded by the drawing specification. In some cases a good-quality steel rule will be adequate; in other cases gauge blocks of a high degree of accuracy may be required. In each case it is the degree of accuracy required and the most efficient way of obtaining it that are the most important considerations.

5.2 Measurement of length

In recent years, the international standard of length has been determined by the use of monochromatic light (i.e. light of a single wavelength) of a precise wavelength. This method has now replaced the old line standard whereby the British legal standard of length, the Imperial Standard Yard, was defined as the distance between two finely engraved lines on gold plugs set in a particular $1''$ square-section bronze bar, and the International Prototype Metre was similarly defined as the distance between two lines on a platinium—iridium bar of a specially designed cross-section. These line standards were kept in London and Paris respectively.

In 1969, the United Kingdom decided to adopt the SI (the international system of units) as its primary system of measurement. With this system the base unit of length is the metre (abbreviation m) which is defined as 1 650 763.73 wavelengths in vacuum of the orange radiation of the krypton-86 isotope. It is not usual to use wavelength standards in the workshop — end standards such as gauge blocks are of far more practical use — however, the accuracy of the gauge blocks is ultimately referred back to the national standard of length.

Gauge blocks

Gauge blocks are the universally accepted standards of length in industry — they are the working standard of linear dimension. They are used for two main purposes:

a) for direct precise measurement where the accuracy of the workpiece demands it,

b) for use with high-magnification comparators, to establish the size of gauge blocks in general use.

Gauge blocks are also used for many other purposes such as checking the accuracy of a measuring instrument or setting up a comparator to a specific dimension, enabling a batch of components to be quickly and accurately checked (the process of comparative measurement is dealt with in section 5.8), or indeed in any situation where there is a need to refer to a standard of known length.

Most gauge blocks are produced from high-grade steel, hardened and stabilised by a heat-treatment process to give a high degree of dimensional stability. Gauge blocks are also manufactured from tungsten carbide, which is an extremely hard and wear-resistant material, although these are initially more expensive than the steel gauge blocks.

Each block has its two opposite measuring faces ground and lapped flat and parallel to within very fine tolerances, fig. 5.1. BS 4311: 1968, 'Metric

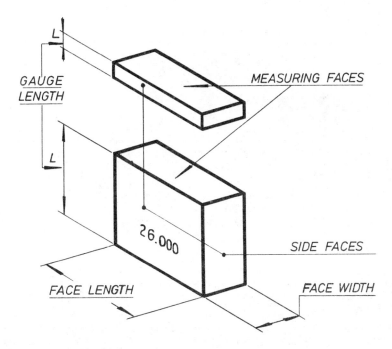

Fig. 5.1 Gauge blocks

75

gauge blocks', gives recommendations covering the manufacture of gauge blocks up to 100 mm in length in five grades of accuracy. The higher grades of accuracy, namely 'calibration grade' and 'grade 00', are used in conjunction with high-magnification comparators as in (b) above; the lower grades of accuracy, namely 'grade 0', 'grade I', and 'grade II', are intended to be used as in (a) above, i.e. for direct measurement where the accuracy of the workpiece demands it.

Grade-0 gauge blocks are used in the workshop for inspection and high-accuracy work.

Grade-I gauge blocks are used for the measurement of components, tools, and gauges.

Grade-II gauge blocks are used in the workshop for rough checks, and for the preliminary setting-up of components where production tolerances are relatively wide.

Gauge blocks are supplied in various sizes of boxed set, fig. 5.2, each block in the set having a different size. The grade of accuracy and the number of blocks in the set chosen by a manufacturer will of course depend upon the type of work for which the set is intended. The three sizes of set detailed below are among those recommended by BS 4311, based upon the type of set that is most in demand, although other sizes of set are available from various gauge-block manufacturers.

Fig. 5.2 Set of gauge blocks

The sets shown opposite are 2 mm based sets — these are recommended as they are less costly and less likely to suffer deterioration in flatness than similar-size series based on 1 mm increments.

Set M 33/2

Range (mm)	Steps (mm)	No. of blocks
2.005	–	1
2.01–2.09	0.01	9
2.10–2.90	0.1	9
1–9	1	9
10, 20, 30	10	3
60	–	1
100	–	1
	Total	33 pieces

Set M 46/2

Range (mm)	Steps (mm)	No. of blocks
2.001–2.009	0.001	9
2.01–2.09	0.01	9
2.1–2.9	0.1	9
1–9	1.0	9
10–100	10.0	10
	Total	46 pieces

Set M 88/2

Range (mm)	Steps (mm)	No. of blocks
1.0005	–	1
2.001–2.009	0.001	9
2.01–2.49	0.01	49
0.5–9.5	0.5	19
10–100	10.0	10
	Total	88 pieces

Gauge-block accuracy   Since gauge blocks are used as 'standards of length' in industry, it is essential to have a set of gauge blocks as a reference set to check that other sets in use are not worn and are still correct in size. In large firms, checking and calibrating is carried out in special rooms, known as standards rooms, which are temperature (20 °C) and humidity controlled. Smaller firms have to send their gauge blocks away to organisations that provide this service.

Care and use of gauge blocks   Gauge blocks are made to a very high degree of accuracy and surface finish, and, if two blocks are twisted together under pressure, it will be found that due to molecular attraction and atmospheric pressure they will adhere firmly together. This process is known as *wringing*, and it is very useful because it enables several gauge blocks to be assembled together to produce a required size. If dirt is present, or if the face of the gauge is damaged, it will not be possible to assemble the gauges together, so there is an in-built checking system.

To determine the gauge blocks required to produce a given dimension, begin with the smallest increment of size and deduct this from the required dimension. Eliminate the next smallest figure in the same way, and continue until the assembly is complete. This procedure will give the minimum number of gauges necessary to build up the given dimension.

Example   To build up gauge blocks to produce a dimension of 69.975 mm:

    69.975
    <u> 2.005</u> 1st block

    67.970
    <u> 2.07 </u> 2nd block

    65.900
    <u> 2.9   </u> 3rd block

    63.000
    <u> 3.0   </u> 4th block

    60.000
    60.0   5th block

Having selected the minimum number of gauges, wipe them clean using a lint-free cloth, a chamois leather, or a cleansing tissue. If the blocks have been protected with Vaseline, this may be removed by a solvent such as petroleum, ether, or carbon tetrachloride.

Begin wringing with the largest sizes first. Avoiding touching the measuring surfaces with fingers (as this can lead to surface tarnish and corrosion) and handling the gauges as little as possible (to avoid errors due to expansion), place two faces together at right angles (fig. 5.3) and, with pressure, twist through 90°. The action should be smooth and with constant pressure − any feeling of roughness will indicate probable dirty or damaged faces and the

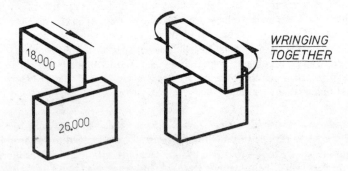

Fig. 5.3 Gauge-block assembly

action must be stopped and the gauge faces examined. When the largest gauges have been assembled, follow with the others in order of decreasing size.

Wringing of gauge blocks should always be carried out over a soft surface, to avoid damage to the blocks if they fall, and gauge blocks should not be left assembled for longer than is necessary. When taking them apart, the gauges are slid from each other (fig. 5.4), cleaned, lightly greased, and immediately replaced in their boxes.

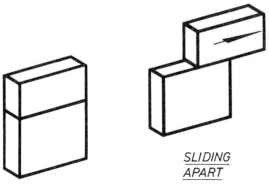

SLIDING APART

Fig. 5.4 Gauge-block dismantling

Two 'protector' gauges are often included in a set of gauge blocks. These are made of the same material as the rest of the set and are 2 mm thick. In use, fig. 5.5, they are placed at each end of the assembled blocks (changing the assembly order as necessary) to ensure that any wear or damage is confined to these two blocks. When they become worn or damaged, replacement is simple and inexpensive.

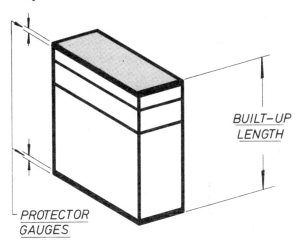

BUILT-UP LENGTH

PROTECTOR GAUGES

Fig. 5.5 Stack of gauge blocks showing protectors

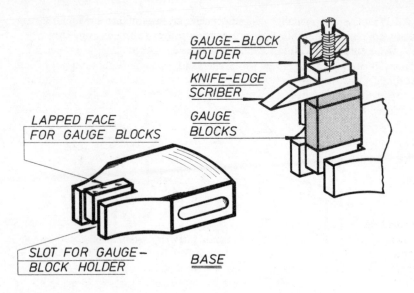

GAUGE-BLOCK HOLDER

KNIFE-EDGE SCRIBER

GAUGE BLOCKS

LAPPED FACE FOR GAUGE BLOCKS

SLOT FOR GAUGE-BLOCK HOLDER

BASE

Fig. 5.6 Base and scriber used as an accurate height gauge

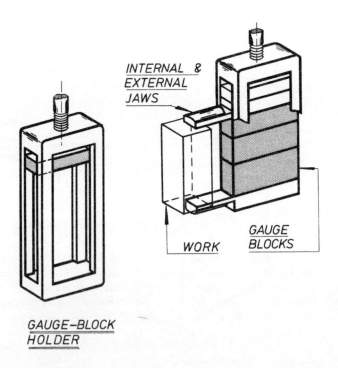

INTERNAL & EXTERNAL JAWS

WORK

GAUGE BLOCKS

GAUGE-BLOCK HOLDER

Fig. 5.7 Jaw gauge blocks and holder used as an accurate external caliper gauge

To increase the versatility of gauge blocks, a range of accessories has been designed to enable various types of precision gauges to be built up. Typical examples are shown in figs. 5.6, 5.7, and 5.8.

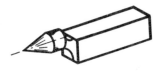

Fig. 5.8 Centre point used for marking out accurate circles and radii

Length bars

When dealing with accurate dimensions that are larger than can be conveniently assembled using gauge blocks, length bars are used, fig. 5.9. Basically, these serve a similar function to gauge blocks on a larger scale. BS 5317: 1976, 'Metric length bars and their accessories', covers bars 22 mm in diameter up to 1.2 metres long, supplied in four grades of accuracy: reference grade, calibration grade, grade 1, and grade 2.

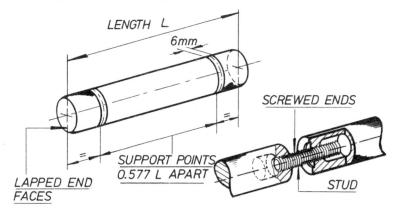

Fig. 5.9 Length bar   Fig. 5.10 Length bars with screwed ends

Reference- and calibration-grade bars have completely plain end faces. Grade 1 and 2 bars have screwed ends, fig. 5.10, allowing a number of bars to be assembled together to a desired length. Length bars are supplied singly or in sets of various sizes together with accessories such as jaws, spherical ends, bases etc. – fig. 5.11.

Airy points   When a length bar is supported horizontally by two supports at its ends, it will sag in the middle. If the supports are moved towards the centre, then the ends will droop. Both of these extremes would lead to the

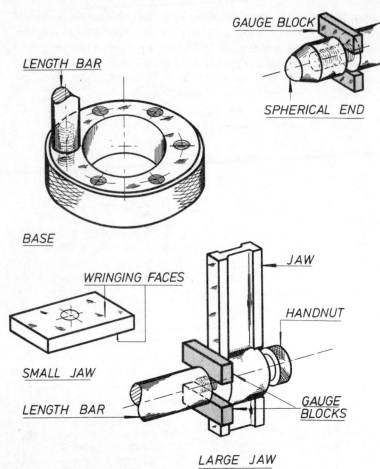

Fig. 5.11 Length-bar accessories

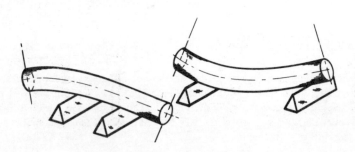

Fig. 5.12 Deflection in length bars — errors in length caused by end faces not being parallel

type of error shown in fig. 5.12. The ideal condition when using a length bar horizontally is to have the end faces parallel. This is achieved by moving the two supports an equal distance from each end to a position where they are 0.577L apart (L = the length of the bar in question), fig. 5.13. These support points are known as the Airy points, after Sir G. B. Airy who first developed the system of support to achieve parallel end faces.

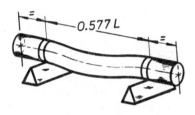

Fig. 5.13 Length bar correctly supported at its Airy points

5.3 Flatness

Flatness is a geometrical quality related to the perfect plane. The method of assessment of flatness is determined by various factors, such as the size and shape of the part and the degree of accuracy specified on the drawing. For very high degrees of accuracy, optical methods beyond the scope of this book would be used, but for general workshop purposes a number of methods are available which will produce satisfactory results.

Surface plates and tables

These are essential items of equipment in most engineering workshops, fig. 5.14. They are the engineer's working standards of flatness and are also widely

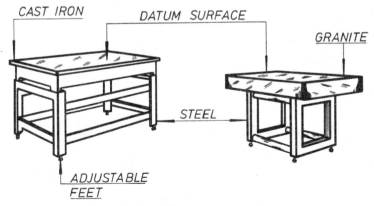

Fig. 5.14 Surface tables

used as datum surfaces. BS 817: 1972, 'Surface plates and tables', recommends that surface plates are cast from a good-quality close-grained plain or alloy cast iron, or made from a close-grained uniform-textured granite.

This British Standard specifies requirements for rectangular and square plates ranging from 160 mm x 100 mm to 2500 mm x 1600 mm, in three grades of accuracy: 'grade AA', 'grade A', and 'grade B'. The grade of accuracy refers to the overall error in flatness of the working surface, and this is defined as the distance separating two parallel planes between which the surface can just be contained, fig. 5.15. The accuracy of the plate will vary according to the size and grade — grade AA is the most accurate and grade B the least.

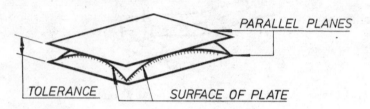

Fig. 5.15 Overall error of flatness

A surface plate 1000 mm x 630 mm would have tolerance errors of flatness of 0.006 mm for an AA plate, 0.012 mm for an A plate, and 0.048 mm for a grade-B plate. In addition to the overall error of flatness, local errors of straightness in the working surfaces of plates 400 mm x 250 mm and larger must not exceed the following in a distance of 100 mm:

    0.0005 mm for grade AA
    0.001   mm for grade A
    0.004   mm for grade B

This is to eliminate the possibility of sudden changes in flatness and yet the surface still lying between the parallel planes, fig. 5.16. Taking the example

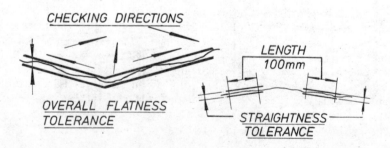

Fig. 5.16 Surface-plate local errors of straightness

of the 1000 mm x 630 mm grade-A plate, the total flatness tolerance is 0.012 mm; however the local deviation in 100 mm is only 0.001 mm.

The surfaces of grade-AA and grade-A cast-iron surface plates are machined and finished by scraping. Grade-B plates can be scraped or machined, and granite surface plates are ground and finished by lapping.

Care must be exercised when using a surface plate. Components or measuring instruments must not be dropped on to the surface, and, when using a surface plate for marking-out purposes, it is not good practice to use a hammer and centre punch on it. When not in use, the surface plate should be wiped clean and protected from corrosion, and the cover should be placed in position to prevent any mechanical damage to the working surface.

As stated earlier, the surface plate is primarily a reference standard for workshop flatness; however, in the workshop it is used more often as a marking-out and inspection table, where the surface is used as a datum, fig. 5.17.

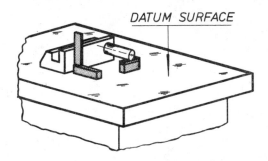

DATUM SURFACE

Fig. 5.17 Surface-plate used as a datum surface

Testing for flatness

If the flatness of a component is to be determined, then a simple test can be carried out using a surface plate as a reference source. The surface of the plate is smeared with a light covering of engineer's blue, and the component to be checked is then carefully rubbed against the plate, fig. 5.18. Any high spots on the component will pick up the blue from the table and become clearly visible, and these can be removed with a scraper. The process can be repeated until an acceptable degree of flatness has been obtained.

A fairly accurate method of determining the proportion of bearing or contact area uses a small glass plate, 50 mm x 50 mm, which is ruled into 400 small squares. This plate, placed on the surface of work that has been blued, is used to assess the fraction in tenths occupied by the high spots in each square. By adding these tenths and dividing the total by four, a percentage bearing area is obtained, fig. 5.19.

BS 308: Part 3: 1972, 'Engineering drawing practice', recommends that permissible errors of flatness are indicated on drawings as shown in fig. 5.20.

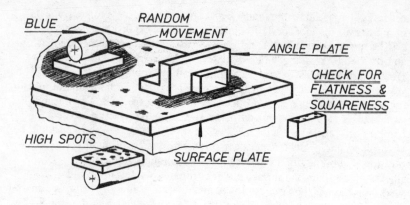

Fig. 5.18 Use of surface plate to check flatness

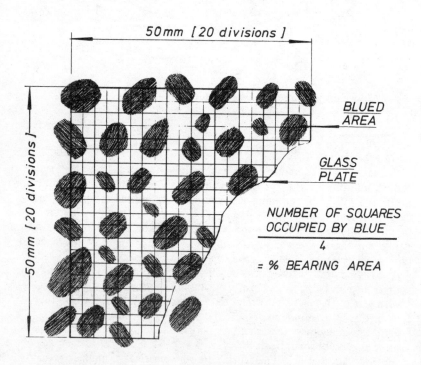

Fig. 5.19 Method of determining percentage of bearing area

Fig. 5.20 BS 308 symbol for flatness

The symbol shown in the left-hand compartment is the symbol for flatness and the figure in the right-hand compartment is the flatness tolerance allowed, i.e. the maximum permissible distance separating two parallel planes between which the surface is contained.

5.4 Straightness

Linear measurement is based upon the concept of straightness. A straight line may be defined as the shortest distance between two points. It is this distance that is important when using such measuring instruments as micrometers, vernier calipers, etc. The geometrical property of straightness is also an important factor that affects the function of many engineering components; for example, the guideways of a planing machine must not only be flat, they must also be straight, fig. 5.21.

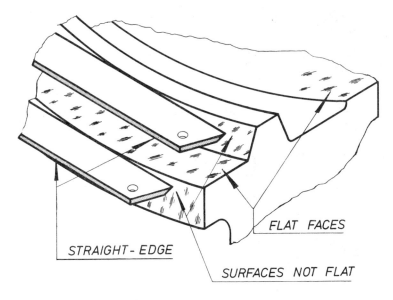

Fig. 5.21 Straightness of planing-machine slideway

The standard for straightness in engineering workshops is the straight-edge. BS 5204: Part 1: 1975, 'Cast-iron straight-edges (bow-shaped and I-section)', recommends two types of design for cast-iron straight-edges:

a) bow-shaped straight-edges for lengths up to 8000 mm, fig. 5.22(a);
b) I-section straight-edges in lengths up to 5000 mm, fig. 5.22(b).

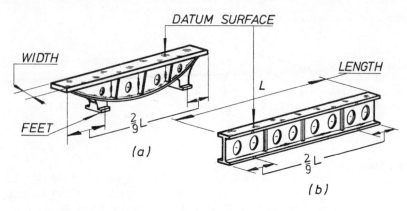

Fig. 5.22 (a) Bow-shaped straight-edge, (b) I-section straight-edge

Part 2 of BS 5204, published in 1977, covers the requirements for steel or granite straight-edges of rectangular section, fig. 5.23, in lengths up to 2000 mm.

Two grades of accuracy are specified for these straight-edges, namely 'grade A' and 'grade B'. These refer to the flatness of the working surface (for practical purposes) and are specified in a similar manner to surface plates, i.e. the flatness tolerance is the distance between two parallel planes that will just contain the working surface, fig. 5.15. As an example, a grade-A cast-iron straight-edge 2000 mm in length would have a flatness tolerance of 0.020 mm; for grade B this would be 0.040 mm. In addition, for cast-iron straight-edges over 1000 mm a local deviation in flatness must not exceed 0.003 mm per 300 mm length for grade A and 0.006 mm per 300 mm length for grade B. For steel and granite straight-edges of 2000 mm length, the flatness tolerance would be 0.015 mm for grade A and 0.030 mm for grade B.

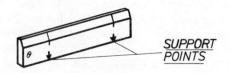

Fig. 5.23 Rectangular-section straight-edge

When using a straight-edge in the workshop to determine the straightness of an engineering component, it may be placed against the work, fig. 5.24, and feeler gauges used to determine any error. Alternatively, a diffused white-light source may be placed behind the straight-edge, so that light will be observed where gaps exist. A gap of 0.000 25 mm can be detected using this method.

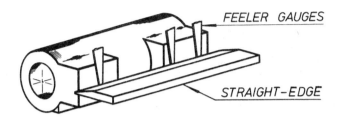

Fig. 5.24 Straight-edge used on its side

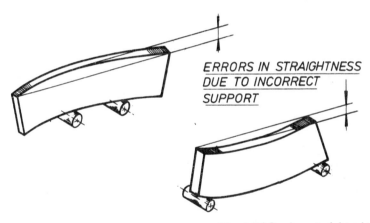

Fig. 5.25 Droop in a straight-edge   **Fig. 5.26** Sag in a straight-edge

When a straight-edge is used on its edge, it should be placed on gauge blocks of equal size, on the surface to be checked, at a distance equal to 2/9 of the length of the straight-edge from each end. The reason for this is that if a straight-edge is supported in the middle it will droop at its ends, fig. 5.25. Likewise, if it is supported at its ends it will sag in the middle, fig. 5.26. If the support points are moved from the middle towards the ends, a position will be reached where the deflection is minimal; this has been found to be $2L/9$ from each end, fig. 5.27. With the straight-edge supported on gauge blocks, the errors of the surface being checked may be determined using other gauge blocks, fig. 5.28.

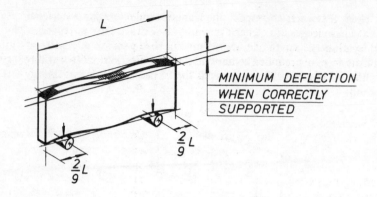

Fig. 5.27 Correctly supported straight-edge

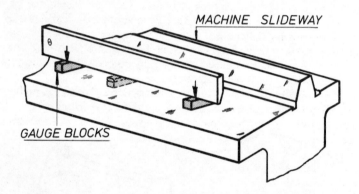

Fig. 5.28 Using a straight-edge to determine the error in a slideway

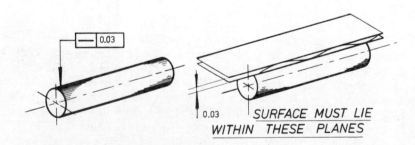

Fig. 5.29 BS 308 symbol for straightness

Note: these support points, which are marked on straight-edges over 1000 mm in length, should not be confused with the Airy points described for length bars. Length bars are required to have parallel ends, whereas straight-edges are required to have minimum deflection of the working surface.

BS 308: Part 3: 1972 recommends that permissible errors of straightness are shown on engineering drawings as in fig. 5.29.

5.5 Squareness

Squareness is a geometric quality that is often taken for granted. A drawing will show two faces of a component at 90° to each other, and it is usually assumed that by producing these faces on a machine tool they will automatically be perpendicular to each other. This is not so — just as it is not possible to produce a given dimension exactly, neither is it possible to produce a perfect right angle. It is possible to get very close to this requirement, but obtaining accuracy is very expensive and the designer must decide just how accurate the workpiece must be to function in the required manner, and must state this clearly on his drawing.

How do we check the squareness of two faces that are required to be perpendicular to each other? One of the commonest methods is to use an engineer's square, fig. 5.30, which is the workshop standard for squareness. By offering up the square to the face being checked, it is possible to estimate the out-of-square condition quite accurately by observing the position and size of the light gaps.

Another method of checking squareness is to use a diffused white-light source. With the stock of the square on a datum face, errors in squareness can be detected by observing the amount and position of light passing between the workpiece and the blade of the square, fig. 5.31.

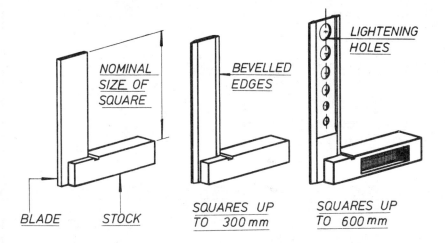

Fig. 5.30 Engineer's squares

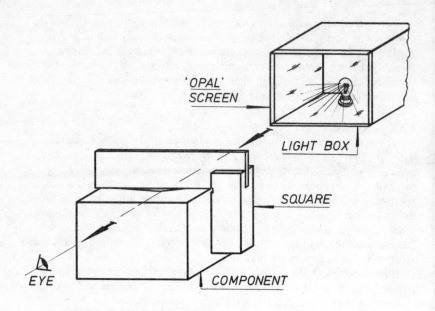

Fig. 5.31 Use of an engineer's square

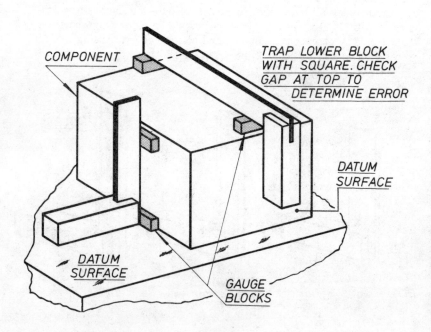

Fig. 5.32 Determining out-of-square condition

92

It is important to note that these methods of checking will indicate only the *existence* of an out-of-square condition. If we wish to determine the value of the error, this is readily obtained by using the square on a datum surface and placing gauge blocks or feeler gauges between the blade of the square and the workpiece, fig. 5.32. A gauge block of known size is trapped between the square and the workpiece at the bottom of the square. Using other gauge blocks, the gap at the top of the square is determined. The difference in size between the two gauge blocks is the out-of-squareness error measured over the length or part of the length of the blade of the square. The direction of the error should also be noted.

Errors in squareness, particularly when dealing with narrow faces, may be checked using an angle plate of known accuracy and a datum surface smeared with engineer's blue. The workpiece is pressed against the angle plate and the face to be checked is rubbed to and fro on the blued surface. A visual indication is then presented of any out-of-square condition – see fig. 5.18.

BS 308: Part 3: 1972 recommends that permissible errors of squareness are shown on drawings as in fig. 5.33.

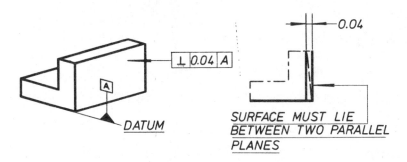

Fig. 5.33 BS 308 symbol for squareness

5.6 Roundness

Roundness is a geometrical quality of form not easily measured. In principle, a 'round' part has all points of its perimeter equidistant from its axis, fig. 5.34. There is some misuse of some geometric terms, although the misuse is often understood. A steel ball and a steel roller are both usually described as 'round'; in fact the ball is spherical and the roller is cylindrical – roundness technically applies to a flat plane, i.e. a circle, whereas in engineering it is more usual to produce a cylindrical shape.

The production of cylindrical shapes can involve some measuring problems – although the diameter is easily measured using linear measuring devices such as the micrometer and vernier caliper, it does not itself indicate the true geometry of the part, i.e. the part may be of the correct diameter but not truly cylindrical as, due to the manufacturing process, the part produced may have become 'lobed'.

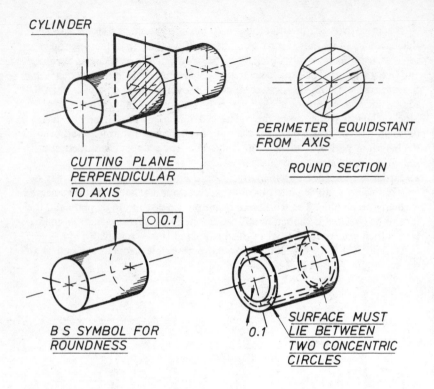

CYLINDER

CUTTING PLANE
PERPENDICULAR
TO AXIS

PERIMETER EQUIDISTANT
FROM AXIS

ROUND SECTION

○ 0.1

B S SYMBOL FOR
ROUNDNESS

0.1

SURFACE MUST
LIE BETWEEN
TWO CONCENTRIC
CIRCLES

Fig. 5.34 Principle of a 'round' part

Lobing

When a component has been produced in a lobed condition, all points on the
perimeter will not be equidistant from its axis, i.e. it will not be round. The
interesting thing regarding lobed parts is that they may be of constant
diameter, e.g. the UK 50p piece (fig. 5.37). Thus, if a circular component is
measured between two fixed positions such as the anvils of a micrometer,
only the size and not the roundness condition is investigated, fig. 5.35.

Usually it is the out-of-roundness condition of a cylindrical component
that is of most interest to the engineer involved in measurement. A method
of determining the amount of 'out of roundness' of a cylindrical component
in the workshop is by the use of a vee block and a dial indicator, fig. 5.36.

5.7 Accuracy of measurement

In any measurement, the possibility of error is bound to arise. No measurement is exact, and the type of instrument used must be chosen to suit the
accuracy of measurement required by the drawing specification. Although a
measuring instrument may theoretically measure to a given accuracy, the
attainment of that accuracy will depend upon several factors. A vernier

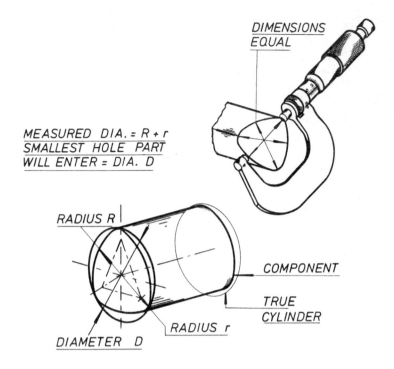

DIMENSIONS EQUAL

MEASURED DIA. = R + r
SMALLEST HOLE PART
WILL ENTER = DIA. D

RADIUS R

COMPONENT

TRUE
CYLINDER

RADIUS r

DIAMETER D

Fig. 5.35 Errors associated with lobed components

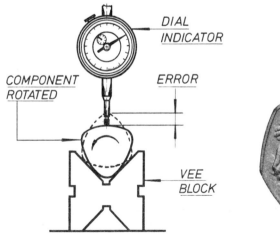

DIAL
INDICATOR

COMPONENT
ROTATED

ERROR

VEE
BLOCK

Fig. 5.36 Method of detecting lobing

Fig. 5.37 Constant-diameter
lobed coin

95

caliper may theoretically read to an accuracy of 0.02 mm, but it would be untrue to say that measurements taken with a vernier caliper may be relied upon to 0.02 mm — the degree of reliability would probably be in the order of ±0.04 mm.

The quality of the measuring instrument (i.e. the accuracy of the scale graduations etc.), the temperature difference between the work and the instrument, and human errors such as sense of feel, parallax errors, and errors due to misalignment of the work and measuring instrument are all factors which may affect the ultimate accuracy of any measurement, and these are discussed below.

It is important not to over-emphasise the effect of any one of the possible sources of error, but to be aware of the capabilities of a measuring instrument or method. The ability to select a suitable procedure for determining a measurement accurately and economically is a basic skill of the technician engineer.

Measurement errors

Some of the possible sources of measurement error are discussed below. It should be noted that more accurate forms of measurement are quite often undertaken in a standards room using accurate and expensive equipment. For the technician in the workshop, however, precise measurement is possible using fairly simple equipment with care and intelligence.

Errors due to temperature variation   Dimensional changes through expansion and contraction due to temperature variations can cause errors in measurement. Generally it may be said that if the measuring instrument and the component are at the same temperature and are of the same material then such errors will be small. If, however, the materials and the temperatures differ, then problems will arise. To overcome these problems, an internationally agreed temperature of 20 °C has been established as the temperature at which all fundamental measurement should take place. Measuring rooms and standards rooms should be kept contantly at this temperature, and humidity and air cleanliness should be controlled.

In the workshop, this degree of control is difficult but, by taking simple precautions, errors can be kept to an acceptable level. For instance, measuring instruments can be kept away from strong sunlight and other heat sources, and work at a raised temperature due to a machining process can be left to cool to room temperature before measurement takes place. Instruments should be handled as little as possible and when not in use should be packed away in protective cases.

Parallax error   This type of error can occur when the line of vision is not directly in line with the measuring scale, fig. 5.38. To avoid this source of error, the line of vision, using only one eye, should be directly over the measuring scale, or over the pointer if using a dial-test indicator. The measuring scale must also be as close to the point of measurement as possible.

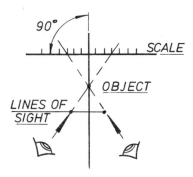

Fig. 5.38 Parallax errors

Layer of lubricant on the component   Most production processes use a cutting lubricant of some description during the machining process. If this lubricant is present in the form of a thin film on a component being measured, the thickness of the film will be included in the measured dimension. Always make sure a component is thoroughly clean before it is measured.

Different inspectors   The development of the correct 'feel' is one of the skills that have to be acquired by the engineer dealing with measurement, and it has been found that, even having acquired this skill, no two people have the same degree of sensitivity. What appears to be normal pressure to one person may be light or heavy to another, and so it is reasonable to expect certain differences in opinion regarding a dimension when more than one person is involved. A way of overcoming this problem is to make use of instruments that apply constant measuring pressure by mechanical means (see 'comparative measurement', section 5.8).

Misalignment   It is under this heading that the greatest number of errors may occur. These errors may be due to inherent inaccuracies of the measuring instrument, or they may be due to the application or method of use of the instrument.

Considering the micrometer, it is known that undue pressure on the instrument may deform the work or the instrument itself, causing error. Also, wear on the anvils and in the micrometer screw will result in inaccurate measurements; for example, anvil faces may wear round, and a ball measured at the outer edges of the anvil will appear smaller than when measured at the centre, fig. 5.39. Again, the anvil faces may be flat but not perpendicular to the axis; consequently a situation could arise where the anvils contact correctly at zero setting, and readings subsequently taken at 0.5 mm intervals would not indicate any error – however, readings taken at full-turn and half-turn increments would immediately indicate error, fig. 5.40. Inaccuracies in the micrometer screw may be checked by measuring a series of gauge blocks and noting the variations.

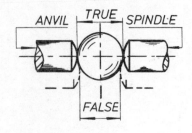 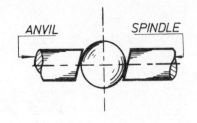

Fig. 5.39 Worn micrometer anvils   **Fig. 5.40** Worn micrometer anvils

Sine and cosine errors occur when the measuring instrument is presented to the workpiece at an angle to the required dimension, fig. 5.41.

It must be emphasised that it is not good practice to rely on the accuracy of the instruments and on the readings taken — readings should be double-checked and the instruments should be periodically checked against the appropriate standards. Measuring instruments are produced to a high degree of accuracy, from the engineer's common rule to the most complex optical instrument, and they should be treated accordingly. Instruments are easily damaged, and very often the damage is not noticeable. Always handle instruments with great care, and report immediately any accidental damage. Protect highly polished surfaces from corrosion by handling them as little as possible and by covering them with petroleum jelly when not in use.

5.8 Comparative measurement

The previously mentioned sources of error in measurement may to some extent be overcome by the technique of comparative measurement, using a comparator. Because a comparator applies a constant measuring force, the influence of 'feel' is eliminated; because less handling is involved, temperature errors are reduced; and because the movement of the comparator's plunger (in contact with the workpiece) is amplified by a suitable system of magnification, more accurate readings are possible. Also, when comparators are used properly, parallax errors are so small that they can be ignored for workshop purposes and, as the measuring plunger is accurately set perpendicular to the work table, errors of alignment are also eliminated.

Once set up (standardised), comparators are relatively simple and quick to use, so they are frequently employed in production processes as an efficient way of checking the accuracy of components as they are produced.

Comparative measurement may be divided into two broad categories:

a) the measurement of length,
b) the measurement of form (geometry).

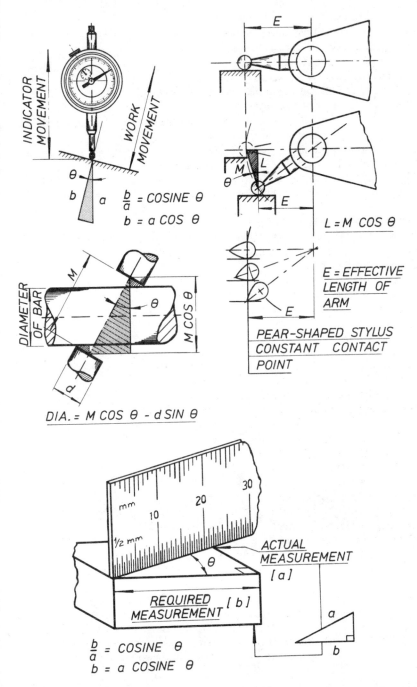

$$\frac{b}{a} = COSINE\ \theta$$
$$b = a\ COS\ \theta$$

$$L = M\ COS\ \theta$$

$$E = EFFECTIVE\ LENGTH\ OF\ ARM$$

PEAR-SHAPED STYLUS CONSTANT CONTACT POINT

$$DIA. = M\ COS\ \theta - d\ SIN\ \theta$$

$$\frac{b}{a} = COSINE\ \theta$$
$$b = a\ COSINE\ \theta$$

Fig. 5.41 Sine and cosine errors

Comparative measurement of length

This is a process of comparing the size of a component with a standard of known size, such as a stack of gauge blocks. It is achieved by using a datum surface and some form of indicating device, the simplest form in general use being the dial-test indicator (DTI) mounted on a strong stand.

The dial-test indicator, is a mechanical device for sensing linear variations. It generally consists of a rack-and-pinion mechanism together with a gear train to give an amplification system such that the scale movement is larger than the plunger movement, fig. 5.42.

In use as a comparator, it is adjusted so that the pointer reads zero when the plunger is in contact with gauge blocks assembled to the required size. The work is then substituted for the gauge blocks and the amount of deviation and its direction are noted, enabling the size of the component to be established.

Comparative measurement of form

There are numerous ways of checking the geometry of a component using comparative measurement. A datum is established and deviations from that datum are shown by the indicator's pointer.

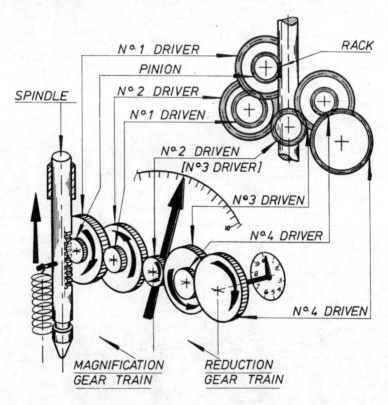

Fig. 5.42 The dial-test indicator

A simple test for roundness is to mount a cylindrical workpiece between inspection centres, with the dial-test indicator plunger in contact with the periphery of the work, as was shown in fig. 5.36. By rotating the workpiece, the out-of-roundness condition will be indicated by the movement of the pointer on the indicator scale. By using a surface plate and a dial-test indicator, checks for parallelism and flatness of a component can be carried out. Figure 5.43 shows a typical comparative method of checking both size and form.

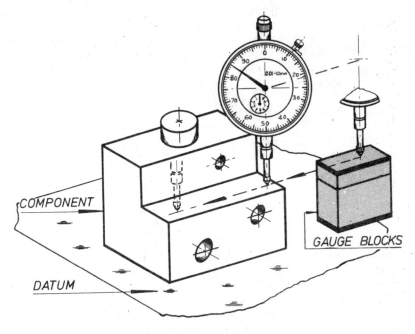

Fig. 5.43 Comparative measurement of size and flatness

5.9 Measurement of angles

A convenient method of measuring angles is to use a dial-test indicator and stand with a measuring accessory called a sine bar. A typical measuring set-up is shown in fig. 5.44.

The important feature of a sine bar is the accurate centre distance of the two identical rollers. By placing gauge blocks of calculated length under one roller, angles may be obtained with great accuracy. Conversely, if a component is placed on the sine bar and gauge blocks are placed under the roller until the component is lying parallel as indicated by the DTI, then, knowing the value of the gauge blocks and the centre distance of the sine bar, the angle of the component may be easily calculated from

$$\text{sine of the angle} = \frac{\text{gauge-block dimension}}{\text{sine-bar centre distance}}$$

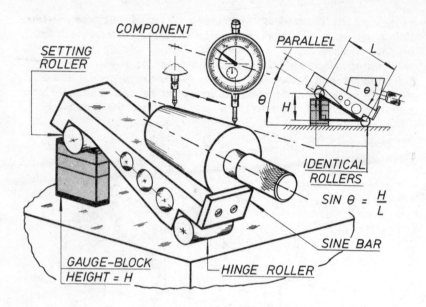

SIN θ = $\frac{H}{L}$

Fig. 5.44 Measurement of angle using a sine bar

5.10 Measuring 'theoretical dimensions'

If the shapes and forms of engineering components are examined, it will be found that they consist mainly of simple geometric shapes such as the cylinder, the cone, and the flat surface, and this means of course that the problem of measurement is relatively straightforward. However, there are numerous cases where measurement is not quite so easy, and measuring techniques have to be developed to overcome the difficulty.

It has already been stated that it is impossible to produce an exact size or angle — it is also impossible to produce a perfectly sharp corner or edge. This means, of course, that a dimension can be shown clearly on a drawing, together with relevant tolerances, but in practice the points indicated do not exist, and therefore a situation arises where a measuring technique has to be devised to measure accurately these 'theoretical dimensions'.

Typical examples of components where 'theoretical dimensions' occur are male and female dovetail slideways, taper plug gauges, taper ring gauges, and other components with taper bores.

Measuring dovetail slides and slideways   The general requirements are that the angular faces of the slides and slideways are identical and the widths are such that a smooth sliding motion exists between the two elements, with no side movement.

First it is necessary to establish that the angular faces are correct, and the most convenient way of doing this is by offering up to the component a

102

master angle gauge smeared with engineer's blue. When it has been established that an adequate bearing area exists (80 % plus), the dimensional measurement may be attempted.

By placing two precision balls or rollers between the angular faces, fig. 5.45, the distance separating them can be readily obtained using gauge blocks, vernier calipers, micrometers, etc. The diameter of the balls or rollers is known, and therefore the centre distance of the balls or rollers can be obtained. By solving the right-angled triangle created by the contact faces of the balls, the theoretical dimension of the corner points is obtained. Finally, by measuring at points along the length of the rollers, a check for parallelism is obtained.

Taper-plug measurement The taper plug shown in fig. 5.46 is set up on end on a datum surface such as a surface plate. Precision rollers supported on gauge blocks are placed in contact, and the measurements M_1 and M_2 are taken. By the use of trigonometry, the angle of taper can be determined.

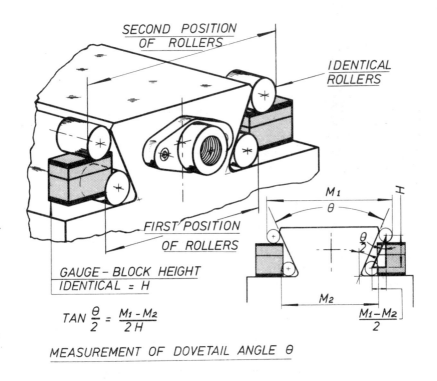

SECOND POSITION OF ROLLERS

IDENTICAL ROLLERS

FIRST POSITION OF ROLLERS

GAUGE – BLOCK HEIGHT IDENTICAL = H

$$TAN \frac{\theta}{2} = \frac{M_1 - M_2}{2H}$$

M_1

M_2

$\frac{M_1 - M_2}{2}$

MEASUREMENT OF DOVETAIL ANGLE θ

Fig. 5.45 Measuring dovetail slides and slideways (*cont'd on page 104*)

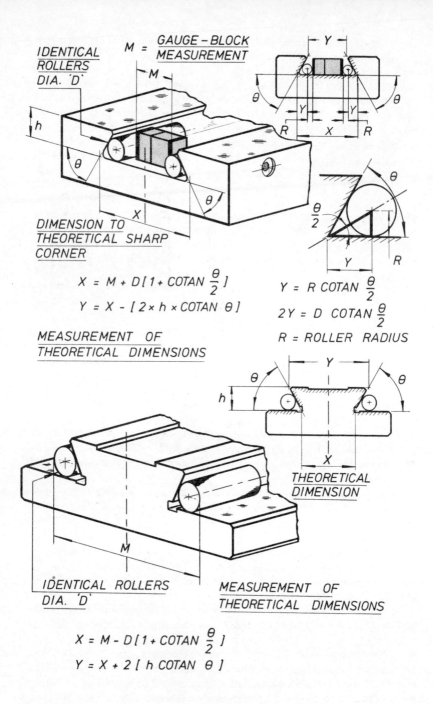

$X = M + D[1 + COTAN \frac{\theta}{2}]$

$Y = X - [2 \times h \times COTAN \theta]$

**MEASUREMENT OF
THEORETICAL DIMENSIONS**

$Y = R \ COTAN \frac{\theta}{2}$

$2Y = D \ COTAN \frac{\theta}{2}$

$R = ROLLER \ RADIUS$

$X = M - D[1 + COTAN \frac{\theta}{2}]$

$Y = X + 2[h \ COTAN \theta]$

Figure 5.45 Measuring dovetail slides and slideways (*cont'd*)

104

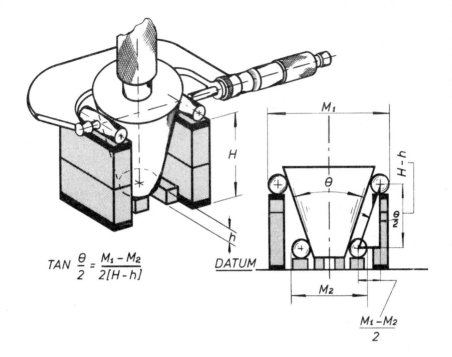

$$TAN \frac{\theta}{2} = \frac{M_1 - M_2}{2[H-h]}$$

Fig. 5.46 Taper-plug angle measurement

Taper-bore measurement   The component is placed on a datum surface, and precision balls of different radii R and r are placed in the bore, fig. 5.47. Measurements M_1 and M_2 are taken and, using the formulae shown, the angle of taper is obtained.

Large taper bores may be measured as shown in fig. 5.48.

Exercises on chapter 5
1   Name the standards used in the workshop (a) for linear dimensions, (b) for flatness, (c) for straightness, and (d) for squareness.
2   Describe with the aid of suitable sketches how each of the standards in question 1 could be used in the workshop.
3   Determine the minimum number of gauge blocks required to obtain the following dimensions, using an M88/2 gauge-block set: (a) 90.5005 mm, (b) 14.999 mm, (c) 150.005 mm.
4   State the procedure for 'wringing' gauge blocks together.
5   Explain the function of a 'datum' surface.
6   When using a straight-edge on its edge, it should be supported on two gauge blocks of equal size. Why is this so, and how are the errors in the component being checked subsequently determined?

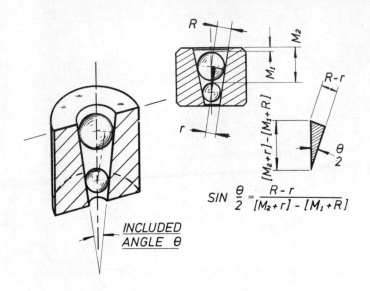

$$SIN \frac{\theta}{2} = \frac{R - r}{[M_2 + r] - [M_1 + R]}$$

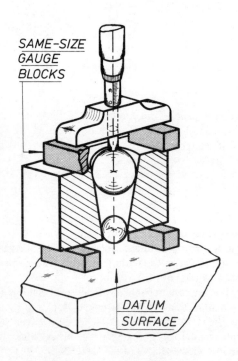

Fig. 5.47 Taper-bore angle measurement

106

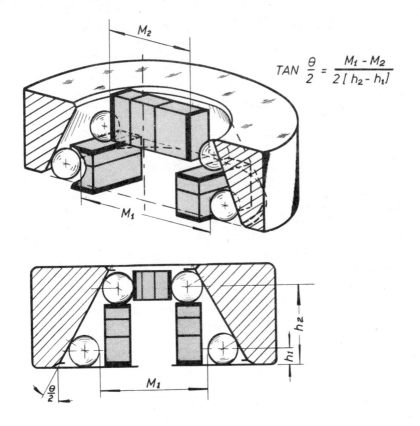

$$TAN \frac{\theta}{2} = \frac{M_1 - M_2}{2[h_2 - h_1]}$$

Fig. 5.48 Measurement of angle of large taper bores

7   Describe with the aid of a sketch how an error of squareness on a component can be within the squareness tolerance given on the drawing.
8   A cylindrical component is required to be straight within 0.015 mm and round within a tolerance of 0.01 mm. Suggest a practical way that this could be checked in the workshop.
9   How does the use of comparative measurement eliminate some of the errors associated with direct measurement?
10   Calculate the size of the gauge blocks required to check an angle of 32° using a 200 mm sine bar.
11   State the factors which can affect the accuracy of the process of measurement.
12   What is the function of a standards room in an engineering organisation?
13   Write a brief statement on the care of measuring instruments.

6   Heat treatment

6.1 Introduction
In other chapters we have looked at the logical order of events that occur in engineering. The processes for producing industry's primary materials such as wire, rods, bars, sheets, plates, castings, and forgings have been outlined. Some of the basic machining processes and the principles of metal cutting have also been described. As well as these processes aimed at producing materials and components, there are other processes that can improve the quality and performance of these products during their working life. Processes such as painting and plating are examples; however, the most important of these processes is heat treatment.

Heat treatment may be defined as the controlled heating and cooling of metal components to produce changes in the physical structure and the mechanical properties of the metal. Heat treatment should not be associated only with hardening, which is but one of many heat-treatment processes: heat treatment may also be carried out to improve the mechanical strength, toughness, machineability, or ductility of a material.

6.2 Constituents and composition of carbon steels
Before dealing with the various heat-treatment processes, we must explain some of the terms used in connection with steel and the changes that occur in it during its heat treatment.

Steel is an alloy of iron and carbon together with small amounts of elements such as manganese, silicon, sulphur, and phosphorus. This material is not regarded as an alloy steel but as a carbon steel (sometimes referred to as a plain- or straight-carbon steel). The properties and structure of a carbon steel are affected mainly by the amount of carbon it contains.

An alloy steel, on the other hand, is produced when the composition of plain-carbon steel is modified by the addition of such elements as nickel, chromium, manganese, tungsten, etc., which gives the finished steel some special property.

Iron
Iron (chemical symbol Fe) is an element which is soft and ductile at room temperatures. It has the latin name *ferrum*; hence products containing quantities of iron are called *ferrous* products.

Iron is what is known as an *allotropic* substance, that is, it can exist in more than one form. At temperatures above 910 °C, the stable form of solid iron has a *face-centred cubic* (f.c.c.) crystal structure; this simply means that its atoms are arranged in a particular pattern, or lattice, and in this form (known as γ-iron, or 'gamma iron') iron can absorb up to 1.7 % carbon. As iron cools below 910 °C, the atoms rearrange themselves into a different stable form with a *body-centred cubic* (b.c.c.) crystal structure; in this new form (α-iron, or 'alpha iron') iron can absorb only very small amounts of carbon — up to 0.03 % maximum, but only 0.006 % at room temperature.

Carbon

Carbon (chemical symbol C) is a non-metallic element contained in all organic materials. Like iron, carbon can also exist in more than one form, for example as a substance known as *graphite*, which is present in cast iron.

Steel does not contain pure carbon or iron as such, but chemically combined compounds of them or solutions of one in the other.

The thermal equilibrium diagram

When cooled from the molten state, steel solidifies at different temperatures, depending upon its carbon content; for example, a low-carbon steel, i.e. a steel with 0.05–0.25 % C, starts to solidify at approximately 1500 °C, whereas a steel containing 0.8 % C starts to solidify at approximately 1450 °C. As cooling of the solid metal proceeds, certain changes take place within the structure, the temperatures at which these changes occur again depending on the carbon content of the steel in question.

These changes can be represented on a *thermal equilibrium diagram*. This type of diagram indicates graphically the relationship between the composition, structure, and temperature of a material.

This chapter is concerned only with the heat treatment of steel and, as previously stated, steel is an alloy of iron and carbon (technically, so are cast irons, which contain iron and carbon in different proportions to those in steel).

By referring to the diagram for the cooling of a carbon steel shown in fig. 6.1, it will be seen that immediately on solidification the hot metal will be in a form known as *austenite.*

Austenite

Austenite is a solid solution of carbon in iron with a face-centred cubic structure. (A solid solution exists when one component of a material is uniformly dispersed in another component of the material in the solid state, and cannot be separated mechanically.) Depending on the temperature, austenite can absorb all the carbon in a steel, i.e. up to the 1.7 % C which will dissolve in f.c.c. iron, and the temperature at which it forms depends on the amount of carbon in the steel. It is important to realise that, just as the solubility of salt in water depends on temperature, so does the solubility of carbon in iron, and in fact at temperatures below 1131 °C austenite will be saturated with less than 1.7 % C.

Austenite is a non-magnetic, soft, malleable material. If, on cooling, austenite becomes saturated with carbon, further cooling will result in carbon being precipitated out of solution in the form of crystals of *cementite* between the austenite crystals.

109

Cementite

Cementite is a chemical compound of iron and carbon, and has the chemical formula Fe_3C. As cementite is a compound, rather than a solution, the percentage of carbon it contains will be constant, and is in fact about 6.7 %. When saturated austenite is slowly cooled, therefore, iron and carbon from the austenite will separate out in a fixed proportion to form cementite, and the remaining austenite will contain a lower percentage of carbon.

Cementite is a very hard, brittle, non-magnetic substance with a whitish appearance. As cementite is so hard, steels containing high percentages of carbon in the form of cementite will also be hard.

Ferrite

If unsaturated austenite containing only a small percentage of carbon is slowly cooled, then the temperature may fall below 910 °C without the austenite becoming saturated with carbon. Below this temperature, the iron begins to change from the face-centred-cubic to a more stable body-centred-cubic form. As we have already said, b.c.c. iron can contain only up to 0.03 % carbon, and a solid solution of up to 0.03 % C in b.c.c. iron is known as *ferrite*. For practical purposes, such as appearance and properties, ferrite may be regarded as being iron, since the small amount of carbon it contains will not have much effect; for example, unlike austenite, ferrite is a magnetic material.

The temperature at which ferrite begins to separate out within the austenite is known as the *upper critical point* (u.c.p.) for the steel, and this again depends on the carbon content of the steel.

Below 723 °C, austenite cannot exist at all but changes to *pearlite*. This temperature is known as the *lower critical point* (l.c.p.) and does *not* depend on the carbon content of the steel.

Pearlite

Pearlite is a mixture of ferrite and cementite, having a laminated structure of alternate layers of dark grey ferrite and whitish cementite, giving a mother-of-pearl appearance under the microscope. The overall percentage of carbon in pearlite is always 0.83 %. Pearlite is a *eutectoid* (a eutectoid is a solid mixture in which changes of structure take place at a single temperature), and austenite containing 0.83 % carbon at or above 723 °C is said to have the eutectoid composition.

After slow cooling to below 723 °C, a steel originally containing less than 0.83 % C will have changed to a mixture of pearlite and ferrite, a steel originally containing more than 0.83 % C will have changed to a mixture of pearlite and cementite, and a steel originally containing exactly 0.83 % C will have changed to the eutectoid pearlite, fig. 6.2.

Martensite

If austenite is not cooled slowly but is instead quenched (i.e. cooled rapidly) then there is no pearlite, ferrite, or cementite formed but instead a substance

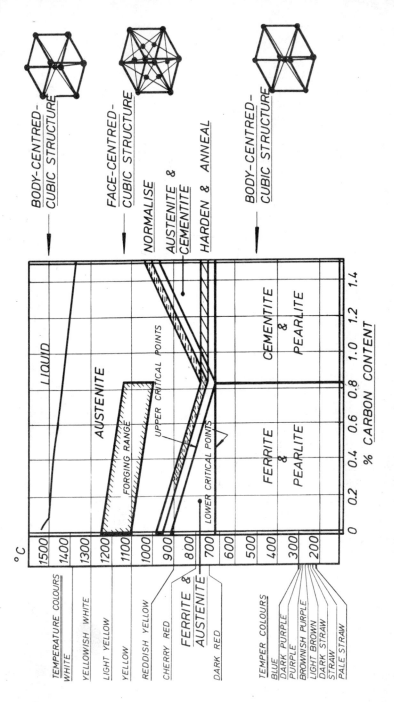

Fig. 6.1 The iron—carbon diagram (general), with atomic structures

111

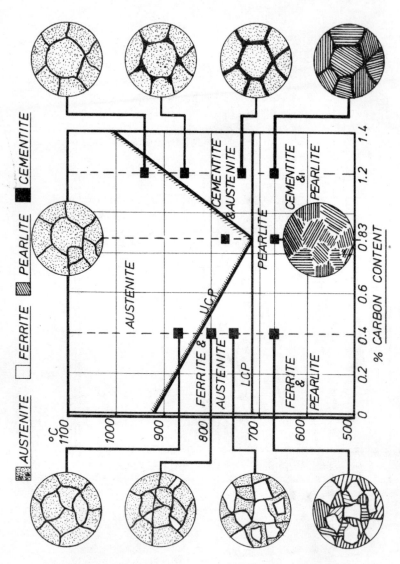

Fig. 6.2 The iron–carbon diagram, with microstructures

AUSTENITE FERRITE PEARLITE CEMENTITE

°C
1100
1000
900
800
700
600
500

AUSTENITE

FERRITE & AUSTENITE

LCP

UCP

PEARLITE

CEMENTITE & AUSTENITE

CEMENTITE & PEARLITE

FERRITE & PEARLITE

0   0.2   0.4   0.6   0.83   1.2   1.4
% CARBON CONTENT

112

Fig. 6.3 Martensite

called martensite, which looks like a mass of needle-shaped crystals if a prepared specimen is viewed under a microscope (fig. 6.3). Martensite is b.c.c. iron supersaturated with carbon, since the rapid cooling has not allowed time for the usual equilibrium composition to be obtained. Because of the supersaturation of the b.c.c. iron, the crystal lattice is distorted and prevents atomic movement. This means that martensite is very hard and strong, and it is the formation of martensite which is initially sought in the hardening of steels.

The amount of martensite formed will depend on the amount of carbon present in the steel and the rate of cooling. If the steel is cooled too slowly, some of the carbon will precipitate out instead of forming martensite, so less hardness will be obtained. The slowest rate of cooling for the maximum amount of martensite to be formed is called the *critical cooling rate* for the particular steel concerned, and this will depend on the carbon content of the steel.

Having now discussed the main terms used to describe the structure of a carbon steel, we can now consider the various heat-treatment processes that are applied to such a steel, and relate these to the iron—carbon equilibrium diagram.

6.3 Annealing

This is a process carried out to obtain steel in its softest state. A steel may be annealed to 'let-down' the hardness or to improve its ability to be cold-worked (i.e. formed in the cold state).

The process consists of heating the steel to its austenitic range, holding at that temperature long enough for the internal changes in the structure to be completed (i.e. from ferrite and pearlite or ferrite and cementite to austenite), then cooling very slowly — preferably in the furnace — to produce the softest pearlitic structure.

Referring to the iron—carbon diagram, fig. 6.1, it will be seen that the temperature required to anneal a carbon steel containing less than 0.83 % C is approximately 50 °C above the upper critical point, whereas the temperature required for carbon steels containing more than 0.83 % C is 50 °C above the lower critical point. The reason for this is that, with steels of higher carbon content, heating to above the upper critical point leads to grain growth (enlarging of the metallic crystals) in the steel, and faults such as cracking may occur if these steels are rehardened.

113

6.4 Normalising

Normalising is a process that is carried out to produce a uniform fine-grained ferrite/pearlite structure. This type of structure is stronger than an annealed structure and is ideal for machining operations. Normalising is quite often carried out on steel forgings, for instance, since holding the steel at the forging temperature for a period of time may result in a coarse structure in the thicker sections. If the temperature of the metal falls during a forging operation, then some of the sections may contain stresses due to mechanical deformation. These faults could lead to distortion of the forging during subsequent operations.

The process of normalising is achieved by heating the steel to its austenitic state, as for annealing, fig. 6.1. The solid solution thus formed is then allowed to cool in air, i.e. at a faster rate than for annealing. This treatment produces a normal fine-grained pearlitic structure free from the defects mentioned above.

6.5 Hardening

Hardening is the process of raising the temperature of a piece of steel until the austenitic structure is obtained, then quenching in a suitable medium. As stated earlier, the rapid cooling action prevents the changes that would normally take place if the steel were cooled slowly (i.e. the transformation from austenite to ferrite and pearlite or pearlite and cementite) and the hard brittle substance known as martensite is produced.

The percentage of carbon present in the steel will affect the amount of martensite produced and hence the final hardness of the steel, fig. 6.4. The amount of martensite that can be produced in a low-carbon steel (0.05–0.25 % C) is insufficient to affect the hardness of the steel to a significant extent, but as the carbon content increases so does the amount of martensite (if the steel is quenched from the correct temperature faster than the minimum cooling rate) and the steel thus becomes harder and stronger. However, we can see from the iron–carbon equilibrium diagram that below the upper critical temperature steels with more than 0.83 % C will contain cementite. This cementite is itself very hard, and if a steel with more than 0.83 % C were quenched from above the u.c.p. it would only become more brittle, i.e. less strong, with very little gain in hardness.

Steels with less than 0.83 % C are hardened by being quenched after heating to approximately 50 °C above the u.c.p. to put all the carbon into solid solution, as can be seen from fig. 6.1.

Steels containing more than 0.83 % C are heated to only 50 °C above the *lower* critical point so that only the pearlite present is converted into austenite before quenching.

6.6 Tempering

A fully hardened carbon steel is extremely hard and brittle; it also contains stresses caused by quenching, due to the uneven rates of contraction (the outside cools and contracts while the inside or core is still in the hot and expanded state).

114

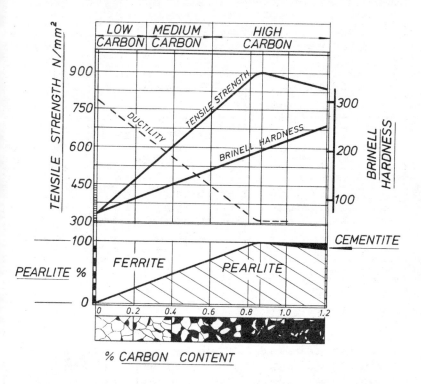

Fig. 6.4 Relationship between carbon content, hardness, strength, and ductility

Tempering is a process that consists of heating the hardened steel to a temperature *below* the lower critical point, causing particles of cementite to be precipitated from the martensite. This relieves the stresses and, although some degree of hardness is sacrificed, the toughness and ductility – and hence the working properties – are improved.

In practice, the amount of hardness required in a steel will depend on the operating conditions of the component in use, and consequently the tempering process will also be varied to produce the required hardness.

Temperatures required for tempering range from 200 to 600 °C and are divided into two groups:

i) low-temperature tempering (200–300 °C) – used where hardness is the main requirement;
ii) high-temperature tempering (400–600 °C) – used where toughness and strength are required.

In the workshop, a common method of judging tempering temperatures for small tools like chisels, fig. 6.5, is to observe the colours produced on the surface of steels when heated. In this method, the steel is polished to produce

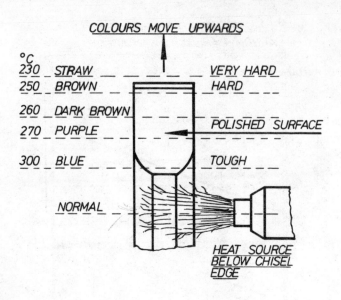

Fig. 6.5 Tempering of a chisel

a bright surface and is then heated. As the temperature rises, the thin oxide film which forms on the surface will assume different colours. When the appropriate colour is visible, the required hardness will have been obtained, fig. 6.6.

Tempering temperature (°C)	Temper colour	Uses
220	Pale straw	Hand scrapers, scribers
230	Straw	Lathe tools, hammer faces
240	Dark straw	Taps and dies, drills
250	Light brown	Woodworking tools
260	Brownish purple	Wood-plane blades, reamers
270	Purple	Press-tools, axes
280	Dark purple	Cold-chisels, screwdrivers
300	Blue	Spring steel, wood saws

Fig. 6.6 Tempering temperature, colour, and use

6.7 Case-hardening
It has already been stated that low-carbon steels (those containing less than 0.25 % C) form very little martensite when heated and quenched, so there is little point in trying to harden this type of material in this manner.

There is, however, a heat-treatment process that enables the lower-carbon steels to be hardened. This process, called case-hardening, gives a dual

116

structure — a tough inner core together with a hard outer case. This is ideal for engineering applications where components are subjected to shock loading but need a hard outer surface that will resist wear. Case-hardening is carried out in two separate stages:

i) carburising the case;
ii) refining the core and hardening the case.

Carburising
This is a process whereby the carbon content or the outer surface of the component is increased to about 0.83 % C by absorption from a substance rich in carbon, fig. 6.7.

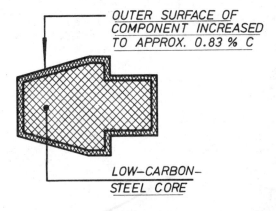

OUTER SURFACE OF COMPONENT INCREASED TO APPROX. 0.83 % C

LOW−CARBON−STEEL CORE

Fig. 6.7 Cross-section of a carburised component

The parts to be carburised are heated to above the u.c.p. in contact with the carburising medium, which may be solid, liquid, or gaseous. The amount of carbon absorbed depends upon the temperature and the length of time the component is in contact with the carbonaceous medium. A case depth of 0.5 mm would be obtained in approximately one hour at 930 °C, but it might take several hours to obtain a case depth of 1 mm at the same temperature. In other words, the rate of absorption is not proportional to time.

Refining the core and hardening the case
It is necessary to refine the core since grain growth will have taken place during the time the steel has been held at the high carburising temperatures. Refining is achieved by reheating the carburised steel to its u.c.p., approximately 920 °C for a 0.25 % C steel, and quenching. Unfortunately, the temperature required for refining the core causes the case to consist of coarse brittle martensite. To obtain a more desirable structure, the case has to be treated again.

117

The carbon content of the case is now approximately 0.83 % C, and by referring to the iron–carbon diagram it will be seen that the upper critical temperature for such a material is 723 °C. By heating to approximately 50 °C above this temperature, the case would be refined, the outer layers of the core would be tempered, and, on quenching, the correct martensitic structure would be formed to give a hard case.

As stated earlier, there are three types of carburising medium: solid, liquid, and gaseous.

Pack carburising This is the process whereby the component to be hardened is placed in a metal box, together with carbon-rich powder or granules. The box should be large enough to hold the parts easily, fig. 6.8, leaving room all round for the carburising agent. Parts may be stacked, provided each layer is well covered with carburising material. The lid of the box should then be sealed with fire-clay and the box placed into a furnace and heated to above the u.c.p. for the material being treated. It is then left to 'soak' for a period of time depending on the depth of case required on the component. On completion, the parts have to be heat-treated to refine the core and harden the case.

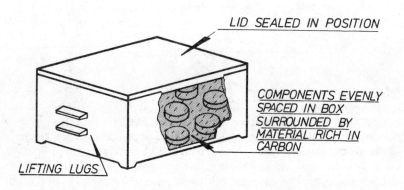

LID SEALED IN POSITION

COMPONENTS EVENLY SPACED IN BOX SURROUNDED BY MATERIAL RICH IN CARBON

LIFTING LUGS

Fig. 6.8 Example of pack carburising

If part of a component must not be hardened, the absorption of carbon must be prevented. For instance, a screw thread would become brittle at the crests of the thread if it were hardened, and it is quite common to leave threads soft. This may be achieved in several ways: the surfaces to be left soft may be copper plated or covered with fire-clay; alternatively, additional metal may be left on during manufacture, and this can be removed after the carburising process, i.e. the carbon-rich layers will be removed during the final stages of manufacture prior to the hardening process, fig. 6.9.

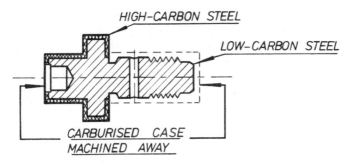

Fig. 6.9 Selective hardening

Liquid carburising This consists of heating the parts to be carburised in a bath of molten salts, the bath usually being a gas- or oil-fired pot made from steel. The salts consist of a mixture of sodium cyanide, barium chloride, and small amounts of other salts. The parts are held at a temperature above the u.c.p. until the desired depth of carburised case is obtained; this is followed by the hardening treatment.

Liquid carburising is widely used for case depths of between 1 and 1½ mm.

Gas carburising With gaseous carburising, the parts to be carburised are heated in a furnace with a controlled atmosphere. The gas used to provide the carbon is a mixture of carbon monoxide, hydrogen, and nitrogen. The parts are again held at above the u.c.p. and left to soak for sufficient time to obtain the required case depth, again followed by the hardening treatment.

6.8 Materials testing
Earlier in this chapter it was stated that heat treatment may be defined as the controlled heating and cooling of metal components to produce changes in the physical structure and mechanical properties of the metal. Mechanical properties are those properties of a material that determine its behaviour when it is subjected to an applied force.

The changes brought about by heat treatment are required so that the material will function in a more satisfactory manner when it is used to produce an engineering component. It is therefore essential to have some methods whereby the mechanical properties of a material may be determined, not only after it has been subjected to one of the heat-treatment processes, but also on receiving the material from a supplier. The commonest tests are those for determining the strength, hardness, toughness, and ductility of engineering materials.

Testing for strength
The strength of a material is its ability to withstand an applied force without failing. As materials react differently to the way they are loaded, it is neces-

119

sary when comparing the strength of materials to state the manner in which they are loaded, as well as the amount of force they can withstand per unit area. Tests for strength may be for tensile, compressive, or shear strength; and the results of the test are usually expressed in MN/m^2 (meganewtons per square metre) or N/mm^2 (newtons per square millimetre), where $1\ MN/m^2 = 1\ N/mm^2$.

The commonest test for strength is the tensile test, carried out on a machine that can apply known forces to a specially prepared test specimen. (These machines are usually of the universal type that, with suitable attachments, can also carry out compression and shear tests.)

While the machine applies the load to the test specimen, the amount of stretch or extension is carefully measured using a device called an extensometer which is attached to the specimen. When the specimen starts to stretch rapidly, the extensometer is removed, to prevent damage, and the load is allowed to increase until fracture occurs. A force—extension diagram can then be drawn and valuable information about the material being tested can be determined from the results of a particular test. This information is essential when a designer has to choose a material for a specific purpose, since undue strength or 'over design' increases the cost of an engineering assembly.

(The results of tests for strength are analysed in engineering-science texts and will not be discussed here.)

Testing for hardness

The hardness of a material is defined as the material's ability to resist indentation and scratching.

Most modern hardness-testing machines measure the resistance of the material being tested to indentation, one of the more widely used methods being that of the *Brinell hardness test*. In this method a hardened steel ball is forced into the material under test by a known load which is maintained for 10 to 15 seconds. After removal of the load, the diameter of the indentation left in the surface of the material is measured in two directions at right angles, using a special low-powered calibrated microscope. The Brinell hardness (HB) is then determined from

$$HB = \frac{load}{surface\ area\ of\ indentation}$$

In actual practice it is unnecessary to calculate the surface area of the indentation left by the ball in the test piece, since tables are available that take into account the size of the ball indenter and the load used to force it into the test-piece surface. Using these tables it is only necessary to know the diameter of the impression left in the test-piece, and the Brinell hardness value is then easily obtained.

In order to obtain accurate and comparable results, the thickness of the test-piece should not be less than eight times the depth of the indentation, and the distance from the centre of the indentation to the edge of the test-piece should be at least three times the diameter of the indentation. British

Standard BS 240: Part 1: 1962, 'Method for Brinell hardness test. Testing of metals', recommends ball indenters of 1, 2, 5, and 10 mm diameter used with varying loads up to 3000 kgf (kilograms force). It is important to select the correct load and ball size for a particular material under test, otherwise the depth of impression obtained may be too shallow or too deep, giving unreliable results.

In order to obtain accurate and consistent results, a ratio of load divided by the ball size squared (load/D^2) is used which, in principle, gives the same hardness value for different loads. Four standard ratios are recommended, namely 1, 5, 10, and 30, for use with the types of material given below.

Material	Load/D^2
Steels and cast iron	30
Copper, copper alloys, aluminium alloys	10
Aluminium	5
Lead, tin, and their alloys	1

If, for example, when using a 10 mm diameter ball and a 3000 kgf load (load/D^2 ratio = 30) it was found that the depth of impression was too deep or too near the edge, then a smaller ball and load giving the same load/D^2 ratio of 30 could be chosen.

As the Brinell hardness test uses a steel ball as an indenter, it is not suitable for testing very hard materials since the ball tends to flatten. It is generally recommended that materials having hardnesses exceeding 500 HB should be tested using a machine with a diamond indenter, such as the Rockwell or Vickers hardness-testing machines.

Testing for ductility
Ductility is the ability of a material to be drawn out or stretched without fracture. Ductility may be expressed as the percentage elongation at fracture of a known gauge length, under tension, and can be one of the properties determined as a result of a tensile test.

The extension of the gauge length (originally 50 mm) is measured by placing the two halves of tensile test-piece together after fracture. The percentage elongation at fracture, which can be taken as the measure of a material's ductility, is given by

$$\text{percentage elongation at fracture} = \frac{\text{extension of gauge length at fracture (mm)} \times 100\%}{\text{original gauge length (mm)}}$$

Testing for toughness
Toughness is a measure of a material's ability to resist shock loading; in other words, the amount of energy a material can withstand before fracture. This is an important property since a component may be required to absorb the kinetic energy of another moving part or to withstand an accidental shock loading without failing. To do this, a material must be not only strong but

121

also ductile, a combination of strength and ductility producing tough materials.

Tests for toughness involve breaking a rigidly held test-piece with a striker attached to a swinging pendulum. One such test is the Charpy test, carried out on an impact testing machine.

Charpy notched-bar test   In this test, a specimen of the material to be tested, 10 mm square x 55 mm long, is notched in the middle of one face with a 45° vee-groove 2 mm deep. The specimen is then simply supported as a beam and the pendulum carrying the striker is allowed to fall under gravity to break the test-piece. The machine is so designed that the pendulum continues its swing, the position it reaches depending upon the amount of energy that was absorbed in breaking the specimen. A pointer on the scale of the machine indicates this value in joules, giving an indication of the material's toughness. The energy on impact can be varied by altering the load on the pendulum to give impact energies of 150 J or 300 J.

6.9 Plain-carbon steels

In the engineering industry, plain-carbon steels are classified in accordance with BS 970. The old identification numbers, the 'En' series, are now being replaced with a six-figure/letter identification, giving the user an indication of the chemical composition and condition of the steel.

Plain-carbon steels are also commonly referred to by their carbon content, and it is generally recognised that there are three types of plain-carbon steel:

a) low-carbon steel, with a carbon content of between 0.05 and 0.25 %;
b) medium-carbon steel, with a carbon content of between 0.25 and 0.6 %;
c) high-carbon steel, with a carbon content of between 0.6 and 1.5 %.

Each of these groups of materials has specific industrial uses, and the choice for a specific purpose of a material from one of these groups will depend on the mechanical properties desired in the finished component and the material's ability to acquire these properties by heat treatment or mechanical working.

Low-carbon steels
It is generally accepted in industry that the low-carbon group of carbon steels may be further subdivided into two categories:

i) dead-mild steel, containing up to about 0.15 % carbon;
ii) mild steel, with a carbon content of between 0.15 and 0.25 %.

Because of their low carbon content, neither of these steels responds to heat treatment.

Dead-mild steel has the characteristics of iron, being soft and ductile, thus it is ideal for parts requiring a lot of cold working and it is widely used for forming motor-car body panels by pressing and for such things as nails, rivets, chains, wire, thin sheet, and solid drawn tubes.

Mild steel is the most widely used of all the ferrous materials, and many thousands of tonnes are used for structural steelwork, for steel plate and sheet for building and industry, and for many thousands of parts subject to low stress, such as nuts and bolts, boxes, bins, rails, shelving, steel cupboards, and forgings.

Medium-carbon steels

As the carbon content of a steel increases, the steel becomes harder and stronger, fig. 6.4; consequently, components that are subjected to more stress are made from medium-carbon rather than low-carbon steel. Such items as crankshafts, machine spindles, drive shafts, gears, small hand tools, and wire rope can economically be produced using medium-carbon steel. This type of steel also responds to heat treatment, and improvements in hardness, toughness, and strength can be obtained by heating and quenching.

High-carbon steels

High-carbon steels can be fully hardened, thus they have an additional range of uses due to this property and are used where hardness and resistance to wear are of prime importance. Punches and dies in press tools, guillotine and bench-knife shear blades, woodworking tools such as chisels and plane blades, cold-chisels, files, and other cutting tools not subjected to high temperatures function extremely well if made from high-carbon steel, when used under the correct conditions. In addition, this steel has many other uses, such as springs, piano wire, and steel cables.

When trying to decide what type of material is to be specified for a particular component, it is necessary to consider (a) the type of conditions under which the material will have to operate, (b) the inherent properties possessed by the available materials, and (c) how these properties may be modified or altered to enable the material the component will be made from to function effectively.

This, of course, is a relatively simple process when considering only one type of material, such as a plain-carbon steel. It becomes much more difficult when a designer has a complete range of ferrous and non-ferrous materials and their alloys to consider.

6.10 Selection of heat-treatment processes

Before selecting a heat-treatment process for a particular component, it is necessary to examine the specification for the component to determine what properties it must possess in its finished state. For example, a cold-chisel made from a plain-carbon steel would have to meet the following specification in order to function satisfactorily:

a) the cutting part (point) must be harder than the material that it will be used on, e.g. low-carbon steel, cast iron, brick, concrete, etc.;

b) the point must be able to withstand severe abrasion but must not be brittle;

c) the body of the chisel must not bend under repeated blows from the hammer;

d) the head of the chisel must be able to withstand repeated blows from the hammer but it must not fracture and cause particles of steel to break away and cause injury to the operator or people in the vicinity.

In the previous section it was stated that high-carbon steels have the advantage of being able to be fully hardened, and it was suggested that the manufacture of a cold-chisel using this type of material would be satisfactory. However, if we look at the specification above, it is obvious that, although this type of steel is strong, it would simply not be suitable in the un-heat-treated state, as the cutting edge and the end of the chisel would crumble under the action of the hammer blows. If, however, we heat-treated the chisel, we could considerably increase the hardness of the point and the chisel end. Unfortunately, when we harden a steel we also increase the brittleness, and in the case of the chisel we would find that, due to this brittleness, the cutting edge and the end of the chisel would shatter when struck with the hammer. Consequently, we would need to further heat-treat the chisel to reduce this extreme hardness. This treatment, known as tempering (p. 114), reduces the hardness slightly and also greatly reduces the brittleness, and our chisel would then be in the desired state of being able to cut other metals and yet would not fracture or crumble when subjected to blows.

6.11 Furnaces
The size and construction of furnaces, and the uses to which they are put, vary considerably. Basically the purpose of a furnace is to raise the temperature of the object put into the furnace. Only the simpler types of furnace will be dealt with here. Many furnaces are specially constructed for continuous production work where a high output is required. The description that follows relates to 'batch' operations, where components are loaded, heated, and then removed.

Non-muffle (open-hearth) furnaces
One of the simplest types of furnace is shown in fig. 6.10. The inside of the furnace is lined with refractory (heat-resistant) bricks, and the heat is supplied either by gas or oil burners or by electrical heating elements built into the walls of the furnace. The work is inserted and removed through the door, which may be hinged, flapped, or of sliding construction.

The temperature obtainable in these furnaces is up to 1500 °C and may be manually or automatically controlled. Unfortunately, with gas-heated furnaces the products of combustion can have a detrimental effect on the surfaces of the work being heated, causing such problems as decarburisation (removal of carbon from the outer layer) of the surface, especially at higher temperatures. Although these problems may to some extent be overcome by adjustment of the air—gas mixture burning in the furnace, there is a design of furnace that isolates the work in a separate chamber, known as a muffle.

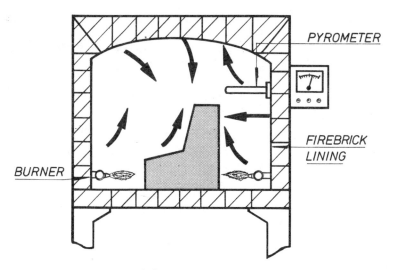

Fig. 6.10 Simple open-hearth furnace

The muffle furnace

A typical gas-heated muffle furnace is shown in fig. 6.11. The work placed in the inner chamber is not in contact with the heat source — the hot gases circulate in the space between the walls. Obviously the size of the work that can be heated is smaller than for a similar size of non-muffle furnace. If neces-

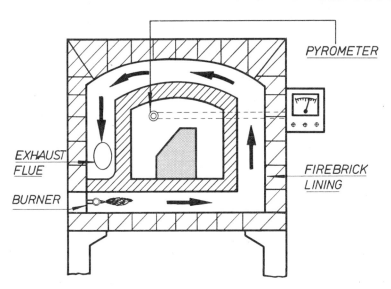

Fig. 6.11 Gas-heated muffle

sary, 'neutral' atmospheres may be introduced into the muffle to protect the parts while they are heated.

With electrically heated furnaces there are no problems with products of combustion, therefore a separate chamber to 'muffle' the work is not required. However, provision is quite often made for the introduction of a 'controlled' atmosphere to the furnace, to prevent oxidation (scaling) etc.

Salt-bath furnaces

These consist essentially of a 'pot' of salts that become molten when heated. Various salt compounds, based upon material like sodium cyanide, chlorides, and nitrates, are available to give treatment temperatures ranging from approximately 150 to 1350 °C.

Figure 6.12 shows the construction of a salt-bath furnace, with the steel outer casing enclosing the heat-insulating refractory bricks and the steel or alloy pot. Salt-bath furnaces can be heated with gas, oil, or electricity. The cowling over the furnace is necessary to remove the obnoxious fumes away from the workshop environment.

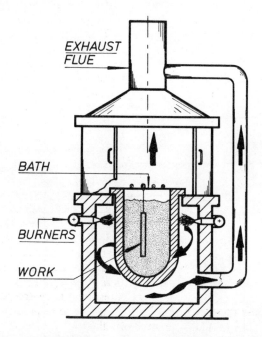

Fig. 6.12 Salt-bath furnace

Care must be taken with loading this type of furnace – it is safest to use a hoist. It is essential to make sure that the components are thoroughly dry, otherwise splattering of the salts will be caused by the rapid formation of steam as wet components contact the hot salts. Cyanide in any form is also an extremely dangerous substance, and advice should be sought before dealing with this material.

6.12 Choice of heating medium

Gas, oil, and electricity are the main sources of heat energy used for furnace heating. Each of these sources has certain limitations and advantages, and the choice of furnace and method of heating will be dictated by the size of the workshop and the heat-treatment requirements — in a small engineering works, for instance, heat-treatment work is contracted out to specialist firms.

Oil The main problem with oil heating is the high initial cost of providing suitable storage facilities and arranging the piping and associated control valves to the furnace. Oil is a reliable source of heat energy and, provided that the burners are cleaned and maintained regularly, will function satisfactorily. Inefficient combustion will cause large soot deposits, and the smell often associated with this system can cause problems. Oil heating is used mainly on the larger type of furnace.

Gas Most factories have gas supplies as part of their main services; connection of a gas furnace to the main supply is therefore relatively simple and can be underground, overhead, or on the surface. There are no problems regarding storage, and control is simple and may be manual or automatic.

Using multiburners, uniform heating over a wide area is possible. By controlling the air/gas ratio, the products of combustion may be varied to give a certain degree of control over the furnace atmosphere, although, as mentioned earlier, the products of combustion may cause decarburisation of the surface of the component being heated.

Electricity This is without doubt the most flexible and convenient source of heat energy: it offers simple operation and control, clean environmental conditions and high efficiency, and is suitable for continuous or intermittent operation. The range of designs available for the heating elements gives great flexibility and compactness in furnace design, and, although operating costs tend to be higher than with gas or oil, initial furnace and installation costs are lower. The heating of the elements is independent of the surrounding air, and therefore no contamination of the furnace atmosphere takes place.

Heating elements are available in a wide range of size, operating temperature, and shape. Elements used for operation in the high-temperature range are made of special high-duty alloys, and these tend to be expensive and their operating life relatively short. Elements often have to be protected from being damaged by work coming into contact with them.

6.13 Measurement of furnace temperature

There are various ways of determining furnace temperatures, including

a) estimating the temperature by judging the colour inside the furnace,

b) marking the job with paints or crayons which change colour at different temperatures,
c) placing in the furnace small cones of various materials that melt and collapse at different temperatures,
d) various types of pyrometer.

Here we shall describe only the now widely used thermocouple pyrometer. The temperature of many modern furnaces is controlled by the use of a thermocouple. This instrument makes use of the fact that if two wires of dissimilar metals are twisted together at one end and a circuit is formed by connecting the other ends to a sensitive voltmeter, fig. 6.13, then, due to the thermoelectric effect, an e.m.f. is produced if the twisted end is heated.

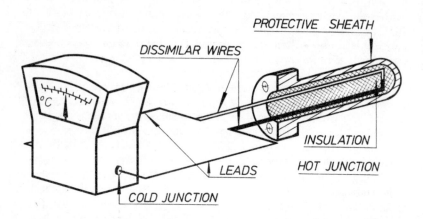

Fig. 6.13 Thermocouple pyrometer

By recalibrating the voltmeter scale (for instance by dipping the twisted ends in boiling water at 100 °C), temperatures may be read directly from the instrument. The twisted end is called the hot junction and the ends connected to the voltmeter the cold junction, and the amount of current flowing in the circuit will depend upon the difference in temperature between these junctions and the types of metal wires used. (Variation in the cold-junction temperature is not usually important in this type of application.)

Typical thermocouple wire combinations are shown in fig. 6.14.

Thermocouple	Temperature range
Copper—constantan	Up to 300 °C
Iron—constantan	Up to 750 °C
Chromel—alumel	Up to 1200 °C
Platinum—platinum/rhodium	Up to 1450 °C

Fig. 6.14 Thermocouple wire combinations

The design of a thermocouple pyrometer is such that it may be conveniently incorporated into a heat-treatment furnace, with the temperature-reading scale placed on the outside, giving direct readings of the furnace temperature at all times (see fig. 6.10). The scale may be at some distance from the furnace, and it is also possible to link the pyrometer with a recording mechanism, so that a permanent record of the furnace temperature over a period of time is shown graphically. Other useful arrangements include the direct connection of the pyrometer to the control valves of the furnace, so that the supply of electricity or fuel can be accurately metered and required temperatures be maintained within very fine limits.

6.14 Quenching

In all heat-treatment processes, the cooling of the heated workpiece is probably the most important and critical part. The rate at which a workpiece is cooled will affect its structure (see 'critical cooling rate', page 113) and hence its hardness, toughness, ductility, etc.

The process of quenching is usually carried out to achieve a sufficiently rapid cooling rate to produce hardness or toughness. The rate of cooling will depend on several factors, such as the cooling medium, the mass of the part being quenched, and the surface area of the component (i.e. the area available for the transfer of heat to the quenching medium).

There are many substances used for quenching, the main ones being *brine*, *water,* and *oil*, in that order of cooling capacity. However, some materials — some alloy steels for instance — have sufficiently low critical cooling rates to allow hardness to be obtained by cooling in an *air blast*, or even in *still air*. In this chapter, though, we shall be concerned only with brine, water, and oil.

Brine will cause the most rapid cooling, followed closely by water. Oil gives a slower rate of cooling. Different rates of cooling are required, as steels with differing compositions need different treatment to obtain different mechanical properties. The most rapid rates of cooling give the greatest hardness; however, there is a danger of distortion and cracking, due to the contractions that take place when a workpiece is quenched. Thus the cooling medium must be chosen to give the best results consistent with these factors.

The manner in which a component is quenched will also affect the finished result. If long slender components were quenched horizontally, the underside layer would rapidly cool, causing contraction and bending. On full immersion the top layers would also cool and try to contract. This would induce stresses into the distorted component and any attempt to straighten the article by mechanical means would probably break it, due to its brittle condition. The correct way to quench such a component is vertically, as shown in fig. 6.15.

If we take a piece of material heated to several hundred degrees Celsius and plunge it into a tank full of brine, water, or oil, it will obviously cool very rapidly. It is important to realise that the heat lost from the component is largely gained by the quenching medium, and it is important to agitate the hot component while it is cooling, so that the desired cooling rate is achieved.

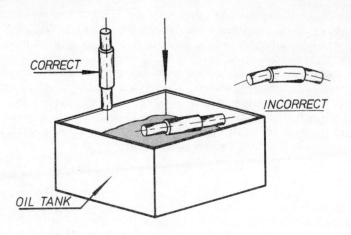

CORRECT

INCORRECT

OIL TANK

Fig. 6.15 Vertical and horizontal quenching

If many articles were to be quenched then it would become necessary to take steps to cool the quenching medium by some means of forced flow to a cooling system.

6.15 Safety

Any heat-treatment process involves the possibility of nearly every known danger: from burns, from lifting heavy loads, of fire, from the products of combustion, from toxic fumes (cyanide), from explosion of liquid salts if not correctly heated or if moisture is present on components, from damaged or exposed heating elements, breakdown of electrical connections, corrosion of furnace casings and work-holding elements, damage to refractory linings, and so on.

It is essential that at all times protective clothing be worn, including suitable overalls, gloves, safety footwear, and safety glasses or visors. Injury can result from the lifting of heavy loads, and when this happens in the vicinity of a furnace or salt bath it becomes doubly dangerous — lifting devices should always be used.

It is important that any heat-treatment area is well ventilated, and furnaces and quenching tanks must be fitted with efficient extraction systems. Gangways and working areas should be kept clear, and work should always be stacked neatly and safely. Manufacturers invariably issue full working instructions regarding the operation and maintenance of equipment and the treatment of the various materials, and these instructions should be carefully followed — if in doubt ask!

The dangers from an outbreak of fire are always present in a heat-treatment area, so fire-fighting equipment must always be to hand and the operation of such equipment must be understood. Air-tight lids should always be fitted to quenching tanks so that any fire occurring can quickly be contained, and these lids should not be removed until the tank or furnace has completely cooled down.

Temperature controls and recording devices are sensitive pieces of equipment and should not be interfered with. For safety's sake, everything should be assumed to be very hot until proved to be cold. Regular checks for gas leaks and of electrical services and earthing are essential, as is regular maintenance of furnaces and refractory linings.

Exercises on chapter 6

1　In a manufacturing process, metals are often subjected to a heat-treatment process. Explain what is meant by heat treatment and give examples of the types of advantage to be gained by such treatment.

2　Plain-carbon steels are divided into three groups. State what these groups are and explain the difference between plain-carbon steels and alloy steels.

3　Make a neat sketch of an iron–carbon diagram up to about 1.5 % C. With reference to the sketch, explain the various changes that take place as plain-carbon steels cool down from the liquid stage.

4　Describe what is meant by the term 'solid solution'.

5　Explain how martensite is obtained and the effect that it has on a steel.

6　State the difference between normalising and annealing. Give examples of where these treatments are applied and the advantages to be gained.

7　A hardening process is often followed by tempering: explain why this is so. Describe the procedure for tempering a fully hardened 0.7 % C plain-carbon-steel cold-chisel.

8　Explain the difference between through-hardening and case-hardening as applied to heat-treated steels. Explain the benefits to be gained by each process.

9　With the aid of sketches, describe the process of carburising using a solid carburising medium.

10　Make neat sketches showing the essential details of a muffle furnace. Explain the advantages of this type of furnace.

11　It is important that the temperature inside a heat-treatment furnace is accurately known and controlled. Explain why this is so and give two methods of measuring furnace temperatures.

12　Quenching (rapid cooling) is an important part of the hardening process. Explain why different types of cooling medium are used.

13　The heat-treatment section of an engineering organisation is one of the more hazardous areas. List the essential safety measures that you would expect to find in operation in that area.

14　Explain the principle of operation of the thermocouple pyrometer.

15　Why is it important to consider the shape of the component when quenching?

16　Describe the factors to be considered when choosing a cooling medium.

7   Plastics

7.1 The uses and nature of plastics

Plastics form an ever-increasing range of materials which permit the engineer to produce a great variety of components. The uses and applications of plastics materials will be found in everyday life, in household kitchen utensils, washing machines, sink units, food mixers, refrigerators, bowls, buckets, food containers, cooking utensils, plastics-topped work areas, etc., to the moulding of quite large ships' hulls. The modern car makes increasing use of plastics materials for seats, body panels, bonnets, radiator grilles, battery cases, steering wheels, distributor housings, electrical insulation, plug leads, air-filter bodies, etc., to name but a few applications.

The building and construction industry also makes use of this type of material for rainwater gutters, pipes, tanks, wall cladding, roofing, electrical plugs and sockets, and cable sheathing. The engineering industry uses plastics in the manufacture of machine parts, gear wheels, brackets, cams, hydraulic and pneumatic hoses, valve bodies, measuring instruments and cases, gauge handles, containers both large and small, electrical panels and components, cable insulation, switches, light fittings, etc. Plastics are also used extensively in the farming industry, hospitals, schools, and other manufacturing industries.

What are plastics and why are they used?

Plastics are synthetic (built-up) materials, and the term 'man-made materials' is often used to describe them. Plastics (for the purpose of this book) are composed of carbon atoms in combination with other elements such as hydrogen, oxygen, nitrogen, and chlorine – they are *organic* materials. (There are also other forms of plastics, for instance those based on the silicon atom, known as *silicones*).

The molecules of an organic substance are held together by strong forces of attraction. However, when the molecules are small (i.e. made up of only a few atoms), the force of attraction between them is relatively small. During the manufacture of a plastics material, a process known as *polymerisation* takes place, as the result of which large molecules are produced, each molecule containing many thousands of atoms in the form of a long chain. As the chains are large, the forces of attraction between them are also large, giving the plastics material its desired properties when fully processed. For example, in the manufacture of p.v.c. (polyvinyl chloride) the gas *ethene* (or *ethylene*, made from naptha, a constituent of crude oil) is heated together with the gas *chlorine* (made by passing an electric current through brine) to produce *dichloroethane* (ethylene dichloride). This is *cracked*, i.e. heated strongly to decompose it, to form another gas called *vinyl chloride*. The simple molecules

132

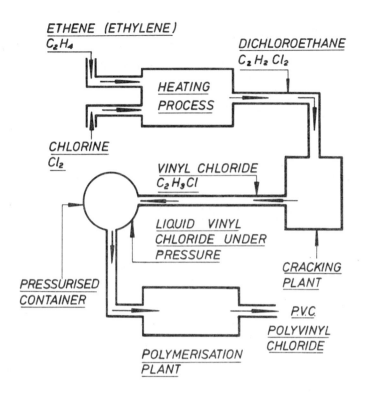

ETHENE (ETHYLENE)
C_2H_4

DICHLOROETHANE
$C_2H_2Cl_2$

HEATING
PROCESS

CHLORINE
Cl_2

VINYL CHLORIDE
C_2H_3Cl

LIQUID VINYL
CHLORIDE UNDER
PRESSURE

CRACKING
PLANT

PRESSURISED
CONTAINER

P.V.C.
POLYVINYL
CHLORIDE

POLYMERISATION
PLANT

Fig. 7.1 Manufacture of p.v.c.

(monomers) of vinyl chloride are then polymerised to form a *polymer* or plastics *polyvinyl chloride*, fig. 7.1.

Plastics are also produced from milk, wood fibres, and other animal and vegetable matter, as well as from crude oil.

Why are plastics used? Plastics possess many advantages over the traditional engineering materials: they are corrosion- and rot-resistant and have good electrical-insulation and heat-insulation properties. They have strength and can be produced in a wide range of colours and effects. Production of plastics components is quick and the cost highly competitive, particularly when large quantities are involved. Plastics are light in weight, have excellent stability, low coefficient of friction, can be produced with a very good finish and appearance, and − often an important feature − can be produced to very close dimensional limits. Against these advantages must be borne in mind the limitations of many plastics, such as an inability to carry very high loads and that, in addition, thermoplastic materials (see section 7.2) tend to lose their strength as the temperature rises above 100 °C.

A wide range of plastics is available to industry, with basic properties varying considerably, and data sheets giving information on mechanical, thermal, and electrical properties are usually available from manufacturers.

Another important feature is that the physical properties of a plastics material may be changed to suit various design requirements. This is done by choosing a basic plastics material and adding such materials as *binders* (to improve rigidity), *fillers* (to give additional strength), *pigments* (to give colour), *plasticisers* (to soften the plastics), *hardening agents*, and *accelerators* (which speed up the hardening process). Thus, if a filler such as asbestos fibre is added to a plastics moulding powder, it will improve the resistance to high temperature, while the addition of mica will increase the electrical resistance.

For the engineering industry, plastics are supplied in the form of powder, rod, sheet, tube, film, granules, syrup, and paste. It is these materials that are further processed to produce an actual component part. The particular plastics material used, and its form, will vary with the type of component to be produced.

7.2 Types of plastics

Plastics materials divide naturally into two distinct groups: *thermoplastic* and *thermosetting* materials. This division greatly influences the design engineer in his choice of plastics materials for a particular component; it also dictates which manufacturing process will be used to produce the component, the production rate, and hence the cost of the article.

Thermoplastic materials

Materials in this group will soften when heated and harden again when cooled. If a thermoplastic material in the form of a powder or granules is heated to a temperature of between 120 and 200 °C, it will soften and flow. If suitable pressure is applied, the softened material can be forced into a mould cavity and will take up the required form. If at this stage the pressure is maintained but the heating is discontinued, i.e. the mould is cooled, the plastics material solidifies and the product is ready for use. If further heat is applied to the product it will again soften and with the application of pressure may be reformed.

This process of softening and forming may be continued a number of times; in other words, during the forming and reforming process, no chemical change takes place which will permanently alter the physical properties of the material.

Properties of typical thermoplastic materials are given in Table 7.1.

Thermosetting materials (thermosets)

Components made with materials in this group cannot be resoftened with heat once they have been produced by the initial moulding process. When a thermosetting material in the form of a powder or resin is heated and becomes softened, pressure is applied and the material takes up the shape of the mould.

134

Due to the heat and pressure, the material solidifies (*cures*), and it is at this point that important changes take place: the physical properties of the material are altered and it is no longer possible for the material to be reheated and softened; it has become quite rigid and cannot be changed in shape or form.

Properties of typical thermosetting materials are given in Table 7.2.

7.3 Plastics-forming processes

There are various ways of producing parts from the plastics materials supplied in the granular, powder, and other forms mentioned earlier. Plastics-forming processes include compression-, transfer-, and injection-moulding, blow-moulding, vacuum-forming, blow-forming, slush-moulding, and rotational moulding. Here we shall consider only compression-moulding, transfer-moulding, and injection-moulding.

Compression-moulding

This was one of the earliest methods of producing plastics components, and it is now the most common technique used for the forming of thermosetting plastics. With compression-moulding, heat and pressure are required to cure the powder or pelleted charge to form a solid permanent component.

A compression-moulding machine consists of (i) a mechanical or hydraulic press with two heated platens to heat the material and to apply the load; (ii) a mould tool consisting of top and bottom halves which, when closed, form a cavity to the shape of the required part. These moulds are located and fixed to the machine and may be changed as the production requirements vary, fig. 7.2.

Moulds can be subdivided into two types: *positive* and *semi-positive*.

Positive moulds (fig. 7.3)   These are the simpler form of mould tool. A cavity in the lower half of the tool forms a well into which the moulding powder is placed. The top of the tool is lowered under pressure and heat is applied to effect the cure. As there is insufficient space to allow any of the moulding powder to escape, all the force of the ram is exerted on the component, whose size and accuracy is affected by the amount of moulding powder in the charge and by such things as temperature and pressure fluctuations. This type of mould can produce dense components with a good surface finish; though, with poorly designed moulds and components, trapped gases generated during the curing process can cause blistering of the surface.

Semi-positive moulds (fig. 7.4)   These types of mould are now widely used, as it is not necessary carefully to measure out the powder charge to obtain an accurate component. The mould is designed to be loaded with an excess of material, which is squeezed out through small clearances in the mould during the curing process. This forms a thin wafer-like leaf of plastic, called 'flash', on the component. This obviously has to be removed, which leaves a small blemish on the surface of the component. This type of mould produces accurate components with uniform wall thicknesses.

Material	Cost	Mechanical properties
Polyethlene (polythene)	Low	Shock resistant but tends to deform under constant load.,
Polypropylene	Moderate	Stronger than polythene; can flex without cracking.
Polystyrene	Low	Reasonably shock resistant.
Polyvinyl chloride	Low	Tough. Impact and abrasion resistant. Rigid or flexible, depending on composition.
Acrilonitrile butadiene styrene (ABS)	Moderate	Shock resistant. Very tough, High tensile strength.
Polymethyl methacrylate (trade name 'Perspex')	Moderate	Tough, strong, and rigid. Shatterproof. Scratches easily.
Polyamides ('nylons')	Fairly high	Very tough. Good abrasion resistance.
Polytetrafluoroethylene (PTFE)	Very high	Low mechanical strength, but properties may be improved by the addition of fillers such as glass, graphite, bronze, etc. Good dimensional stability.

Table 7.1   Properties of typical thermoplastic materials

Other features	Max. service temperature	Typical uses
Easily moulded. Resistant to chemicals. Good electrical insulator. Obtainable in various colours, translucent or opaque. Available in low- and high-density forms.	90 °C	Containers, bottles, bowls, buckets, toys, packaging, flexible tube, waterproof sheeting.
Can be subjected to temperatures above 100 °C, thus useful where sterilisation is required. Extremely light (relative density 0.90).	120 °C	Electrical components, car parts, distributor caps, fans, containers.
Easily processed. Very light and resistant to moisture. Glossy translucent finish. May be 'expanded'. One of the most widely used plastics.	80 °C	Low-temperature applications such as refrigerator liners and containers. Expanded, as ceiling tiles and packaging material. Also for thin-walled cartons such as those used for cream, yoghurt, etc.
Good electrical insulator. Water and weather resistant. Obtainable in various colours, transparent or opaque. Non-flammable.	90 °C	Rigid: rain-water guttering and downpipes. Flexible: items previously made from rubber, e.g. shower curtains, rainwear, garden hoses, flooring, artificial leather.
Resistant to chemicals. May be chromium-plated.	115 °C	Safety helmets; (plated) auto parts, radiator grilles; (laminated) car body panels.
Easily machined and bent. Very good light transmission. Obtainable coloured or as a clear glass-like material.	90 °C	Dials, handles, knobs, nameplates, control panels, windows, machine guards, roof lights. Suitable for both inside and outside use.
Low coefficient of friction. Wax-like. Good heat resistance. Dimensional stability affected by water absorption.	120 °C	Gear wheels, bearings, door catches, small mechanisms, nx nuts, bolts, locking inserts clothing (as a textile fibre), ropes, tennis-racquet strings, fishing line.
Lowest coefficient of friction of all plastics. Good heat and chemical resistance. Will not absorb moisture. Difficult to injection-mould.	250 °C	Bridge bearing pads, electrical insulation, oil seals, surgical heart-replacement parts, valves, sealing tapes, high-pressure gas hose, general anti-friction purposes, non-stick surfaces on cooking utensils.

Material	Cost	Mechanical properties
Phenol-formaldehyde	Low	A resin produced when phenol and the gas formaldehyde are heated together. Good surface hardness. Brittle when compression moulded.
Urea-formaldehyde	Low	A clear resin syrup produced when urea and formaldehyde are heated together. Powder produced from the syrup can be compression moulded. Hard and rigid with good resistance to scratching. Harder than phenol-formaldehydes.
Melamine-formaldehyde	Moderate	One of the hardest of common plastics.
Epoxy resins	Moderate	Great strength and toughness when in laminated form.
Polyester resins	Moderate	Usually used in conjunction with glass fibres to produce a tough material with a very good strength/weight ratio.

Table 7.2 Properties of typical thermosetting materials

Other features	Max. service temperature	Typical uses
Good dimensional stability. Generally used with 'fillers' such as cotton or fabric cuttings, or woodflour (very fine sawdust) added to the resin. Other fillers include mica (for improved electrical resistance), asbestos (improved heat-resistance), and glass. A limited range of colours, usually dark.	120 °C	As moulded parts such as utensil handles, fuse-box covers, meter cases, toilet seats, instrument panels, etc.
Fillers added to improve properties, as with phenol-formaldehyde. Lower heat- and water-resistance than phenol-formaldehydes. Good resistance to chemicals and solvents. A wide range of colours may be produced.	80 °C	Decorative material, e.g. on cars and furniture. Adhesives, bottle tops, knobs, caps, etc.
More resistant to heat and water than urea-formaldehydes, and lighter colours can be obtained. Decorative laminates obtained by impregnating patterned paper with the resin.	130 °C	Laminates used for table and work tops ('Formica', 'Warevite', etc.), control panels, cups, saucers, plates, etc.
Cured by the addition of a hardener at room temperature (heat speeds up the process). Excellent resistance to solvents, alkalis, and some acids. Able to bond to other materials. High electrical resistance.	200 °C	Adhesives. 'Potting' of electrical components, i.e. allowing the resin to set around a component to keep out moisture, oil, dirt, etc. and hence increase the life of the component.
Good electrical and corrosion resistance. Good sound-damping properties.	95 °C	Used with glass fibres to produce laminates for boat hulls, car bodies, etc.

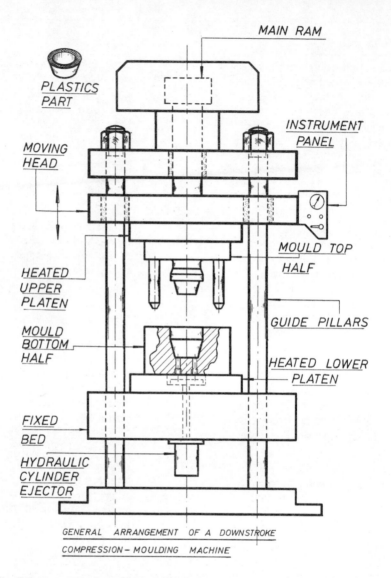

MAIN RAM

PLASTICS PART

INSTRUMENT PANEL

MOVING HEAD

MOULD TOP HALF

HEATED UPPER PLATEN

GUIDE PILLARS

MOULD BOTTOM HALF

HEATED LOWER PLATEN

FIXED BED

HYDRAULIC CYLINDER EJECTOR

GENERAL ARRANGEMENT OF A DOWNSTROKE COMPRESSION—MOULDING MACHINE

Fig. 7.2 A compression-moulding machine

The disadvantage with this method is that more moulding material is used, although the flash may be recycled by grinding it up and using a proportion of it as a bulk filler. Also, additional handling time is required to remove the flash from the component.

With both types of mould, the cavity is filled with the moulding material, the two halves are closed, and heat and pressure are applied and maintained for a

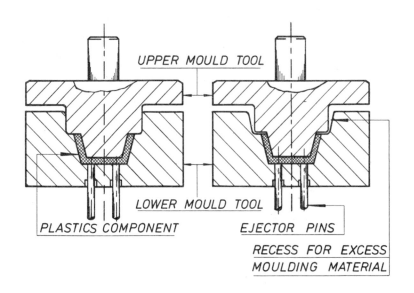

Fig. 7.3 A positive mould   **Fig. 7.4** A semi-positive mould

given period of time, during which the material solidifies (cures). After this stage the mould is opened and the component is ejected.

Typical moulding temperatures vary between 140 and 190 °C, and production costs are generally low. Large components may be produced by compression moulding, although the cycle time is rather slow, thus fewer components are produced each hour so the cost of each article produced is proportionally higher. It is difficult to produce intricate shapes and hold dimensions to very close limits using this process.

Transfer-moulding
Where a component is of intricate shape, or must not have flash, a process known as transfer moulding is used, fig. 7.5. With this method, the mould is designed with a separate charging or heating chamber and, when fluid, the charge is forced by a plunger into the mould cavity through small channels known as *gates*, enabling even the most intricate moulds to be completely filled. Heat and pressure are maintained to effect the curing. When cured, the component plus the runners and sprue is removed from the mould. The runners and sprue are subsequently removed from the component, fig. 7.6.

Injection-moulding
This is probably the most widely used process for producing plastics components. It may be used for all of the thermoplastic range of materials, and it is also one of the fastest processes for producing components of a wide range of sizes and shapes. In this process the material is heated to obtain

141

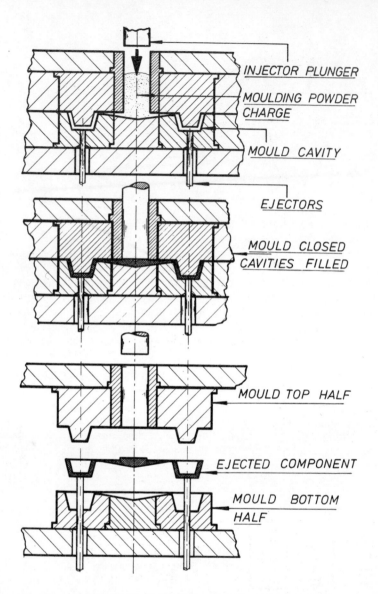

INJECTOR PLUNGER

MOULDING POWDER
CHARGE

MOULD CAVITY

EJECTORS

MOULD CLOSED
CAVITIES FILLED

MOULD TOP HALF

EJECTED COMPONENT

MOULD BOTTOM
HALF

Fig. 7.5 A transfer mould

fluidity and is then forced into the mould cavity, where it is allowed to solidify.

An injection-moulding machine has three main elements: (1) a hopper and injection cylinder, (2) a heating zone, and (3) a moulding area, fig. 7.7(a).

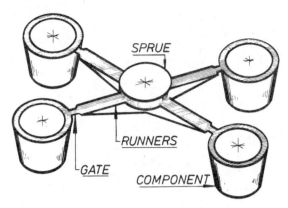

Fig. 7.6 Runners and sprue

In operation the required amount of plastics material is metered from the hopper into the injection cylinder, where a plunger or screw, figs 7.8 and 7.9, forces the material into the heating zone (where the temperature is raised under controlled conditions). At the same time, a similar quantity of softened material is displaced under pressure through a nozzle in the cylinder end into the mould cavity, until the mould is full. Pressure on the material is maintained until it has solidified, which is often speeded up by the use of water circulating around the mould tool. The mould is opened, the component is ejected, and the process is repeated. Production time is rapid, being as short as a few seconds for small components.

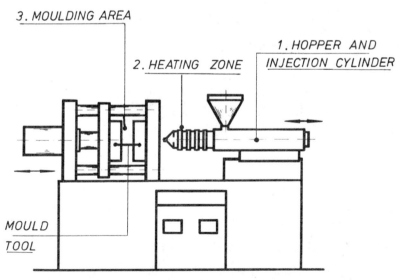

Fig. 7.7(a) An injection-moulding machine

143

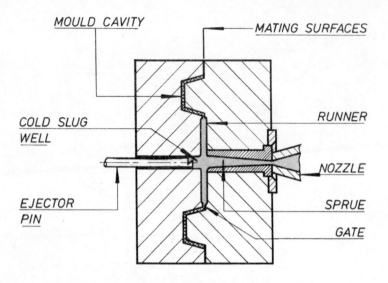

Fig. 7.7(b) Section through a typical injection-moulding tool

The mould tool is located and fixed to the machine and, as with the compression moulds, is changed as demands for different components arise.

The heating area of the machine consists of a chamber enclosed by electrical heating elements. The actual design of the chamber is such that the material is forced to the outer wall by the 'torpedo', to obtain more uniform heating.

Components produced by the injection-moulding process vary in mass from a few grams to several kilograms. They can be dimensionally accurate (within ±0.025 mm) and of complex form. They may be produced in a wide variety of thermoplastic materials and in a wide range of colours. The cycle time (i.e. the time to produce one component) can be very short, particularly with automatic or semi-automatic machine operation and using multi-cavity mould tools.

7.4 Advantages and limitations of moulding processes

The moulding processes discussed in the previous section are very versatile and, with well designed moulding tools, the engineer can produce components that are in the finished state and ready for immediate use, or at least in a state that requires very little extra work to complete them.

A disadvantage of injection-moulding is that the mould tools are very expensive to manufacture and the initial tooling cost is not fully justified unless the component is required in very large numbers.

Components produced by compression-moulding tend to be of the larger type — it is difficult to produce components of an intricate form or with thin

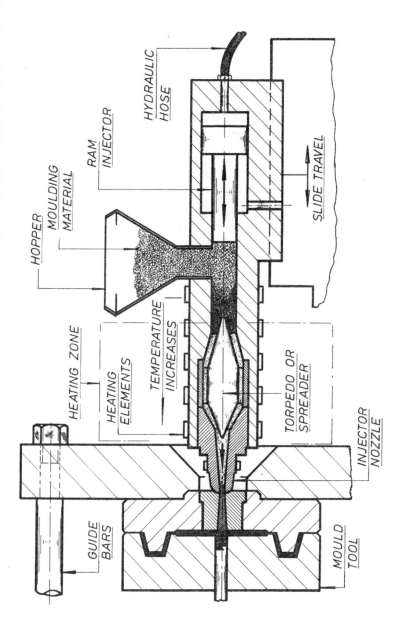

Fig. 7.8 Injection plunger

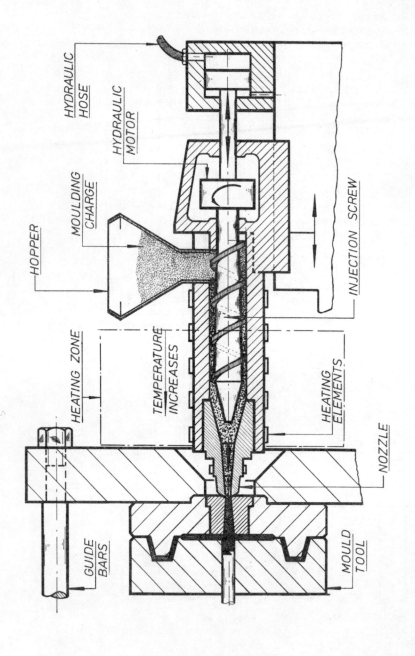

HYDRAULIC HOSE

HYDRAULIC MOTOR

MOULDING CHARGE

HOPPER

INJECTION SCREW

HEATING ZONE

TEMPERATURE INCREASES

HEATING ELEMENTS

NOZZLE

GUIDE BARS

MOULD TOOL

Fig. 7.9 Injection screw

146

sections — and, since there is a greater time lag while the material solidifies, production times are slower. Generally the cost of the mould tools is lower than that for injection-moulding tools.

Whichever type of process is used, a plastics component should not merely be considered as a part-for-part replacement of a metal one, but rather as a component where the advantages of a plastics material and its production processes are fully utilised. For example, the ease with which plastics may be formed (formability) enables the engineer to produce components which it would be extremely difficult if not impossible to produce in steel or similar material. In turn, this enables the designer to design a component so that a number of separate parts are incorporated into a single plastics component, this often leading to an overall reduction in size, cost, mass, and production time. An example of this is shown by the design of the box in fig. 7.10, where the tray, the lid, and the hinge are all moulded as a one-piece component. It has been claimed that replacing several metal components with a single plastics one has in some cases resulted in a reduction in cost of 95 %, but this sort of figure would be exceptional rather than a general rule.

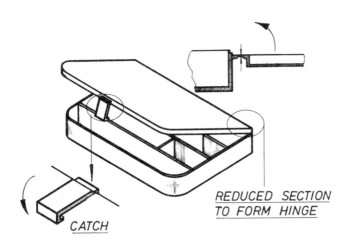

REDUCED SECTION
TO FORM HINGE

CATCH

Fig. 7.10 Example of designing in plastics

Plastics materials are easily machined, and a component is sometimes moulded to an approximate shape and then machined to a finished shape to give additional accuracy of size or geometry (e.g. flatness).

7.5 The use of inserts
It was stated earlier that components produced by the compression- and injection-moulding processes require little or no finishing; however, there are certain instances where a plastics component may be subjected to higher than normal loading at specific areas, and at these areas additional strength is

required. This strength can be obtained by using metal inserts at the required points.

Screw threads are notoriously weak in plastics, and they can also be very difficult to mould. To overcome this inherent weakness, threaded steel or brass inserts can be incorporated at the moulding stage or fixed in position later. With thermosetting materials, metal inserts are generally moulded in position, and this means that the insert has to be placed in position in the mould tool before moulding; thus production time is longer. Also, great care must be taken to ensure that the insert is not dislodged during moulding.

Thermoplastic materials can have the insert positioned after the moulding operation is completed. One method is to press into the moulding an insert which has been previously heated. This melts the plastics locally and, as the insert cools, the plastics also cools, solidifying around the insert. Other methods are to assemble the insert with ultrasonic tools applying high-frequency vibrations, or by using the type of insert shown in fig. 7.11. In this example, a hole is moulded in the normal manner or is drilled in position. The threaded metal insert is pressed into the hole, causing the body of the insert to close, and at this stage the parts are assembled and the screw fitted. This action forces the insert body open, pressing part of the metal into the plastics itself, and the insert cannot then be withdrawn.

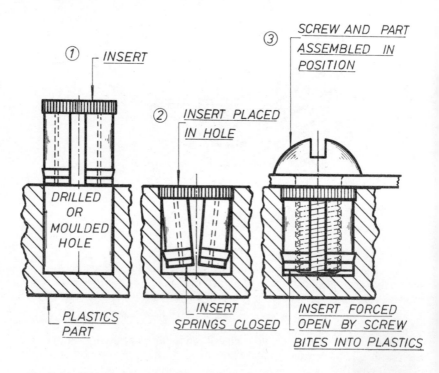

Fig. 7.11 Threaded metal inserts

Wherever possible, components should be designed so that the inserts can be moulded in position – this saves second-operation work and eliminates handling time. In practice, the insert is supported by the mould tool during the moulding process, and the insert itself will be designed with flats, squares, undercuts, and knurled surfaces to ensure that it is firmly held in the plastics component and cannot turn or be pulled out, figs 7.12 and 7.13.

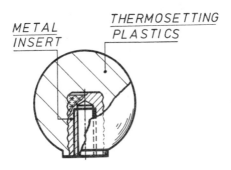

Fig. 7.12 A moulded-in insert

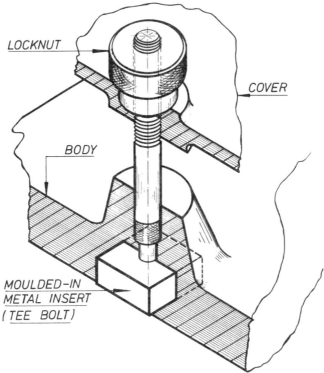

Fig. 7.13 Moulded-in tee bolt

7.6 Choice of material and process

When choosing a plastics material for the production of a component, a number of factors must be considered.

Is the part to be moulded or to be machined from standard forms of material supply such as bars, tube, rod, thick sheet, etc.? This will depend entirely on whether the part is required in sufficient quantities to justify the expense of mould tools, or whether it would be more economical to machine the part from standard material. For instance, a gear wheel produced in large quantities from nylon would be injection-moulded; on the other hand, the same gear with only small numbers required could be machined from nylon or laminated plastics material using normal machine-shop equipment.

It was stated, in section 7.2 that plastics divide naturally into two distinct groups — thermosetting and thermoplastic. Which group is chosen? Various factors must be considered: will the component be subjected to high temperatures? will it have to resist strong forces acting on it? is it of intricate shape? does the component have to operate outside and be subjected to all weather conditions? does it require metal inserts to be moulded in position? will machining operations be required to complete the part? and so on.

At this stage it is important to note that, when designing a plastics component to replace a metal one, the physical properties of the component need not be exactly the same. For example, considering a replacement design for a household steel bucket or bowl, the chosen material could be polypropylene (thermoplastic). The plastics item would be softer and more flexible than the original one, and yet the overall result would be an improvement, i.e. lighter, resistant to corrosion and chemical attack, more attractive (variety of colours), would not damage other surfaces, and cheaper. Another example is the replacement of a metal container by a plastics one, fig. 7.10. Again, the rigid steel is replaced by a softer material and the advantages are the same as in the previous example, with the added advantages that one moulded component replaces four metal ones, there are no assembly problems, and the firm's name and other information can be moulded on the lid or box.

Example 1   A plastics gear wheel for the tumbler-reverse mechanism on a small lathe is required (a) to transmit light loads, (b) to be reasonably accurate in tooth form, though not precisely so, (c) to be self-lubricating, and (d) to be quiet running. What material should it be made from?

Answer: A nylon (thermoplastic) moulded gear would fulfil these requirements, with low cost and high production rates.

Example 2   A ball handle for fixing to a gear-change lever requires (a) to be rigid, for handling purposes, (b) to be capable of receiving a metal insert, and (c) to have instructions, numbers, or directions moulded on it. Its mass is not important, but variation in colour would be an advantage. What material should it be made from?

Answer: In this case a thermosetting material such as urea-formaldehyde could be used.

150

Choosing the plastics material for the component will generally govern which moulding process is used, as most thermosetting materials are compression-moulded and thermoplastics are injection-moulded. The nylon gear would be injection-moulded using a multi-cavity tool; a bucket or bowl would be injection-moulded using single-impression mould tools. An instrument panel or casing would need to be rigid and possess strength (to protect the instruments) and so a compression-moulded thermosetting material would be chosen. A ball handle, again rigid, and with a metal insert, would be transfer-moulded, to ensure that the mould is completed filled.

7.7 Safety

As stated earlier in this chapter, most plastics require heat to soften or melt them and pressure to move the melt into the mould tool. The combination of heat and pressure can be dangerous, especially as most of the plastics involved have a softening temperature above the boiling point of water. Great care must be taken when using plastics-moulding and -forming machinery. Operators must be supplied with protective clothing, i.e. a long-sleeved overall, heat-resistant gloves, safety shoes, etc. — these will give protection from the very hot surfaces in and around the mould and machine, and also from the hot moulding itself, which may at times reach temperatures of around 240 °C. Safety glasses are also essential, to give protection against chips of flying plastics. Great care must be taken when using hygroscopic (water-absorbing) materials — these must be completely dried out before use, otherwise steam will be formed in the mould cavity, giving rise to dangerous situations.

Mould tools must be accurately positioned and fastened before starting the moulding operation. Mould faces must be kept perfectly clean and free from any obstructions. The mould tool must be completely closed before the moulding cycle begins. The moulding machine itself must be completely guarded in such a manner that hands or fingers cannot be trapped when the two halves of the mould come together. Recommended moulding temperatures must be carefully regulated, as excessive temperatures can lead to plastics materials decomposing and giving off obnoxious and dangerous gases. Moulding areas must be well ventilated, floors kept clean, and gangways kept unobstructed. Completed components should be loaded in pallets or boxes and kept neatly stacked.

Exercises on chapter 7

1 A plastics handwheel is required for a control valve in a hot-water system. Describe in brief concise terms the design considerations necessary before large-scale production. Consider such things as (a) the type of plastic to be used, (b) the name of the plastic, (c) advantages and disadvantages, (d) obtaining the desired colour and finish, (e) physical weaknesses and how to overcome these, (f) the moulding process used to produce the handwheel.

2 State four hazards and dangerous conditions existing in a factory producing plastics. Explain how the machine operator can be protected against these dangers.

3　The design of a mould tool will depend on the type of plastics material being used. Name and briefly describe three different types of mould.

4　Make a neat sketch of a plastics mould, name four parts of the mould, and explain the function of each.

5　'Thermosetting' and 'thermoplastics' are two important names given to plastics material. Explain what is meant by these names and suggest a possible moulding process to be used for both types.

6　Explain how mechanical weaknesses in moulded plastics components may be overcome by introducing metal into the design of the finished component. Name the metal part involved and describe how it is incorporated into the finished component. Name two kinds of metal used for this purpose.

7　Make a neat sketch of two different plastics components, showing clearly the areas where additional strength is obtained by including metal parts in the finished design.

8　Explain the meaning of injection-, compression-, and transfer-moulding and show by neat sketches the essential differences between the three processes. Name the type of plastics material associated with each process.

9　State the advantages and disadvantages normally associated with injection-, compression-, and transfer-moulding processes.

10　Plastics components have replaced a large number of items previously constructed of metal. Name six replacement designs and discuss the reasons for using plastics materials and the advantages gained by so doing.

11　Plastics are often described as 'synthetic' materials. What is meant by this term? What is an 'organic' material?

12　Explain the purpose and reason for using binders, fillers, pigments, plasticisers, and accelerators with plastics materials. Name two materials used to give improved electrical and heat-resistance to a plastics material.

13　Thermoplastic material may be in the 'rigid' or 'non-rigid' form. Explain what is meant by these two terms and give two examples of each. State the advantage and disadvantage of each type.

14　What is the meaning of the term 'curing' as applied to plastics materials? If curing is ineffective, what will be the result?

15　A small transmission gear is to be produced from plastics material. State two different materials suitable and describe the manufacturing processes used for each.

16　Give two advantages and two disadvantages of using plastics materials for producing gear wheels.

17　Choose a suitable plastics material for producing a door handle. Describe the material and make a neat sketch showing any inserts necessary. Describe the moulding process associated with the chosen plastics material.

18　Name four articles of clothing produced from plastics materials.

19　Name four machine-tool parts made from plastics materials.

Index

Airy points, 81
annealing, 113
austenite, 109

billets, 3, 9
blast furnace, 2
blooms, 8, 9

capstan lathes, 62
carbon, 109
carbon steel, 108, 122–3
carburising, 117–19
case-hardening, 116–17
cast iron, 4
cementite, 109, 110
chip formation, 37–9
cogging mill, 8
comparative measurement, 98–101
compression moulding, 134, 140
continuous casting, 4, 9
contraction allowance, 6
cope, 6, 7
copying, 61
core, 6, 7
core print, 6, 7
critical cooling rate, 113
cupola, 5
cutting fluids, 49–52
    application, 52
cutting forces, 43
    measurement, 43–5
cutting tools, 33
    carbide-tipped, 48
    materials, 47
    negative and positive rake angles,
        40, 41
    normal rake angles, 34

single-point, 33–6
tool-in-hand system, 34
tool-in-use system, 35, 42

degrees of freedom, 59
die, 11
    drawing, 10
    extrusion, 11
    press tool, 21
drag, 6, 7
drilling machine, 64
    operations, 65
ductility, 10
dynamometers, 43–5

equilibrium diagram, iron–carbon,
        109, 111, 112
extrusion, 11
    cold, 12
    materials suitable for, 11

facing sand, 6
ferrite, 110
fettling, 8
flatness, 83–7
    testing for, 85
fly press, 19
forging, 12
    grain flow, 12, 13
    operations, 14
forming, 61, 73
furnaces,
    heating method, 127
    muffle, 125
    open-hearth, 124
    salt-bath, 126

furnaces (*cont'd*)
temperature measurement, 127

gauge blocks, 75
accessories, 80
accuracy, 77
care and use, 77
wringing, 78
generating, 61, 72
grain flow
forging, 12, 13
presswork, 28
green sand, 6
grinding machine, 67
cylindrical, 68
surface, 68

hardening, 114
heat treatment, 108

in-gate, 6
ingots, 3
injection moulding, 141, 144
inserts, 147
plastics moulding, 148, 149
iron, 108
iron—carbon equilibrium diagram,
109, 111, 112
iron ore, 1
iron and steel products, 9

lathes, 62—4
length bars, 81
accessories, 82
Airy points, 81
lobing, 93, 94

machine tools, 54
degrees of freedom, 60
elements, 55—9
generating, forming, and copying,
61

malleability, 10
martensite, 110, 113
materials testing
ductility, 121
hardness, 120
strength, 119
toughness, 121
measurement, 74—105
accuracy, 94
of angle, 101
comparative, 98
errors of, 96, 97
of length, 74—83
of theoretical dimensions, 102
metal cutting
analysis, 36
shear plane, 37
types of chip, 38, 39
milling machine, 69
dividing head, 71
horizontal, 70
rotary table, 71
vertical, 70
moulding box, 6
moulding flask, 6
moulds
injection, 144
positive, 135, 141
semi-positive, 135, 141
transfer, 141, 142

normalising, 114

oblique cutting, 40
orthogonal cutting, 39
out-of-roundness, 94

parallax, 96
parting sand, 6
pattern, 6
pearlite, 110
pig iron, 3
plasticity, 11
plastics, 132

154

choice of, 150, 151
manufacture, 133
mould tools, 135
types of, 134, 136–9
power press, 20
press tools, 21
  blanking layouts, 30
  clearance, 24
  follow-on tools, 22
  operations, 26
  punch and die, 21
  shear, 24
presswork, 18
  costs, 18
  fly press, 19
  power press, 20
primary forming processes, 1, 4
  advantages and limitations, 15
pyrometers, 127–8

quenching media, 129

rapping, 8
riser, 6, 7
rolling, 8, 10
roundness, 93
  out-of-roundness, 94
runners, 6, 7

safety, 16, 31, 53, 130, 151

sand casting, 6
shaping machine, 66
  operations, 67
sine bar, 102
sine and cosine errors, 99
sinter, 2
slabbing mill, 8
slabs, 3, 9
slip gauges, *see* gauge blocks
solid solution, 109
squareness, 91
  checking, 91–3
steel
  carbon, 46
  high-speed, 46
straight-edges, 88
  minimum deflection, 89
straightness, 87, 90
surface plates and tables, 83
  testing for flatness with, 85

tempering, 114–16
thermocouples, 128
transfer moulding, 141
tungsten carbide
  cutting tools, 48
  dies, 11
turret lathes, 63
tuyères, 3

wire drawing, 10, 11